WOLFGANG SMITH FOR EVERYONE

WOLFGANG SMITH FOR EVERYONE

Science, Metaphysics, and Religion

WITH TESTIMONIALS AND TRIBUTES

Edited by
Bruno Bérard and Marie-José Jolivet

Angelico Press

First published in French as
Wolfgang Smith pour tous: Introduction à la physique—
Et témoignages-hommages à la suite de sa disparition

For information, address:
Angelico Press, Ltd.
169 Monitor St.
Brooklyn, NY 11222
www.angelicopress.com

ppr 979-8-89280-204-8
cloth 979-8-89280-205-5
ebook 979-8-89280-206-2

Book and cover design
by Michael Schrauzer

Table of Contents

Preface

This book was intended as a tribute to Wolfgang Smith on his 95th birthday; published posthumously, it remains a tribute, coupled with an introduction to his work, particularly to physics stripped of all scientism.

Wolfgang was a friend since 2008, and Marie-José Jolivet is the translator of his more recent works. That being said, the following texts are the work of each contributor in their own name.

The fact that Wolfgang Smith was a mathematician, physicist, and metaphysician is rare enough to be worth mentioning at the outset. There have certainly been others—Plato, Aristotle, Descartes, Leibniz—but in this age of scientific overspecialization, a cross-disciplinary approach is most welcome and valuable.

While one of the characteristics of Wolfgang's writings is that they are clear and precise—"to the point," as we would say in English—it might seem surprising to consider this presentation. However, as we shall see, it was not useless to provide an overview of the many subjects covered: quantum physics, perception, cosmology, evolutionism, Teilhardism... We will then discover a coherent thought process, a Weltanschauung—a worldview—capable not only of offering a way out for physics, which is trapped in an impasse, but above all of providing an understanding of the world and of man, under the gaze of an irrefutable transcendence.

Of course, the aim here is not to develop a thesis on the life and work of Wolfgang Smith; however, the following elements should provide an overview and facilitate understanding—without failing to refer to the master's books.

Bruno Bérard

Introduction

Wolfgang Smith and Initiation

When exactly I'm not sure, but at some moment in my own intellectual development I discovered a group of thinkers who revealed to me the deep initiatory dimension of philosophical enquiry. This dimension, of course, was central to philosophy's origins, but was lost concurrently with—and not unrelatedly to—the discipline's secularization.

The thinkers I have in mind are Stratford Caldecott, Jean Borella, Wolfgang Smith, and some others. These men were connected in at least two ways: they had all been profoundly influenced by the French thinker and critic of modernity René Guénon and other twentieth-century "Traditionalists" or "Perennialists"; and they had each brought new life into Catholic discourse by integrating that prior learning into their Christian thought in marvelously creative ways. Moreover, they had all had their writings published by Sophia Perennis and Angelico Press.

Smith, on whom the present volume focuses, especially captured my attention due to the sheer *wholeness* of his project. For there are philosophers who devote their lives to one question, constantly developing and tweaking their answer to it. There are also philosophers who devote their lives to many questions and never seem to offer satisfactory answers to any of them. But Smith stands among a tiny number of thinkers throughout the history of mankind who answered many questions and did so comprehensively. And Smith not only did so with remarkable depth, but he presented to his readership how his manifold insights fitted together as a unified and coherent account of reality—that, of course, is what we really mean, or ought to mean, by *wisdom*.

Smith's was not, though, a "system." A philosophical "system" implies conceptual ownership—or worse, conquest—of the reality of which that system purports to give an account. Smith, on the other hand, remained a humble beggar at the feet of Lady Sophia his entire life. Hence, his works do not offer us a schematic structure by which we might reduce the world to an assemblage of abstractions. Quite the reverse; his thought breaks us free from the chamber of unanchored abstractions and by stages initiates us back into the corporeal world of our experience, indicating all the implications of such an initiation along the way.[1]

Were I (perhaps foolishly) to attempt a summary of Smith's key contribution to our civilization's intellectual culture, I would say that contribution was threefold: 1) a new, richer dialogue between, and reconciliation of, modern science and classical philosophy (occasioning the overcoming of reductionist materialism); 2) a creative dialogue between the Catholic intellectual tradition and the great spiritual and philosophical traditions of the world; and 3) a retrieval of gnosis as integral to the Christian *Weltanschauung*.

By presenting the major themes of his corpus in this way, one can see that "wholeness" is exactly what characterized Smith's scholarship: wholeness in cosmology; wholeness in human wisdom, encompassing all truth wherever it is found; and wholeness with regard to the union of man with his Creator.

As is well known among his readership, in developing his case Smith challenged the largely unscrutinized and thus prejudicial option in favor of a lingering Cartesian bifurcationism, which he identified as the false epistemology enabling many physicists to posit the quantum realm and the corporeal realm as coextensive within the same horizontal plane of reality. He challenged this paradigm by retrieving the ancient notion of a tripartite structure to all being, a structure found in all

[1] See Wolfgang Smith, *The Vertical Ascent: From Particles to the Tripartite Cosmos and Beyond* (Philos-Sophia Initiative Foundation, 2020), 59–70.

traditional religions and wisdom traditions, albeit presented diversely.

What implicitly—and sometimes explicitly—ran through all his writings was the notion that investigation into the nature of any given level of reality entailed a moral response on the part of the enquirer. And in this regard, I insist that his writings are in strict continuity with the ancient philosophical tradition in being essentially *initiatory*.

The physical world—the world explored by the physicist—reveals through mathematical and experimental investigation a realm of indeterminacy, in which *materia prima* is being called forth—or actualized—by the reception of quality. In turn, the physical world reveals the power of the Creator in the created order, as He brings into being the corporeal world of our experience, which in its actualization reveals to us something of the inner life of its Author. The world is, then, truly theophany.[2] And effecting the actualization of this corporeal world of our experience *is* the spiritual world, or aeviternal world, which is vertically contiguous with—and thus causally related to—the corporeal world. Hence, intellectual and spiritual enquiry at each level leads to some intimation of God and His self-revelation in the cosmos, pointing us in different ways to the pursuit of divine wisdom, or, to adopt Smith's preferred term, *gnosis*.

But how does gnosis differ from, say, catechesis? Smith makes a remark (almost in passing) in a little-known essay entitled "Science and Myth: The Hidden Connection":

> It is not the function of doctrine to take us out of the founding myth: to "explain it away." On the contrary, its function is to bring us into the myth; for indeed, the pearl of truth resides in myth as in a sanctuary. Authentic doctrine can take us to the threshold of that sanctuary; but like Moses before the Promised Land, it cannot enter there.[3]

[2] *The Vertical Ascent*, 140.

[3] http://www.worldwisdom.com/public/viewpdf/default.aspx?article-title=Science_and_Myth--The_Hidden_Connection_by_Wolfgang_Smith.pdf.

This is probably the pithiest presentation I have ever encountered of the distinction—of especial importance to Clement of Alexandria—between the pistic Christian and the gnostic Christian. Here, "gnostic" is not meant in the pejorative sense of Manicheanism and its antecedents and successors in the Western world. Indeed, against such phony "gnosis" Smith railed.[4] Rather it is a gnosis at which Smith himself arrived firstly by the rejection of two types of dualism.

The first dualism was that of the above-mentioned Cartesian bifurcationism that long ensorcelled the Western mind and continues to undermine credence in the world of our experience, reducing "reality" to quantity and measurement. The second dualism was older, namely that of "double-truth" that posits supposed truths of rational and scientific enquiry to be unrelated to—and likely in conflict with—the truths of religious belief and spiritual experience.

Smith's response to such dualisms was that of placing scientific findings within a framework of classical participation metaphysics, combining the ontology of the Platonic and Neoplatonic traditions with the corporeal realism of the Thomistic tradition. The result has entailed a renewed and serious dialogue between contemporary science and classical philosophy, a retrieval of our world after a centuries-long dominance of epistemological skepticism, and a recovery of the theophanic cosmos as a ladder of being by which to ascend to spiritual transformation through union with God.

Consequently, Smith's achievement also entails the recapture of *personal* meaning and purpose, for no one can hope for a meaningful and purposive life in a meaningless and purposeless universe. Hence, inadvertently, Smith has rejuvenated the discipline of philosophy as a pathway of initiation, which is what it always should have been. What I refer to in this last point, I have witnessed unfolding before my eyes

4 *The Vertical Ascent*, 137–54.

as I have introduced my students to Smith's teachings since being appointed to the Wolfgang Smith Chair in Philosophy at St Mary's University, London. I am immensely proud to hold a chair named in his honor. It is clear to me that his mission has only just begun.

Dr Sebastian Morello
Wolfgang Smith Chair
St Mary's University,
Twickenham, London

September 29, 2025
Michaelmas

Testimonials & Tributes

I met Wolfgang Smith in the late 1990s in a large bookstore in New York. I had gone in with the intention of finding the latest news on astrophysics, but my gaze fell on a book whose title caught my attention: *Teilhardism and the New Religion, a thorough analysis of the teachings of Pierre Teilhard de Chardin.* During the 1950s, I had experienced a surge of enthusiasm for the writings of the honorable Jesuit, but it had quickly faded and never returned. I sensed a certain critical tone in the expression "*New Religion*" that prompted me to take the book home with me. Curiously enough, the back cover said nothing about the author, who was completely unknown to me. It was only after reading his biography on the last page that I realized he was not a theologian, but a scientist of the highest caliber. The indictment of Teilhard de Chardin seemed to me to be judiciously balanced by a presentation rich in references to Scripture and to the Christian tradition as I had received it. This is what struck me as the originality of the work: critical, certainly, but restoring for the reader what Catholic dogma has taught for centuries. It was then that I decided to translate it and, in doing so, to broaden my knowledge of the author by reading his works available at the time.

Today, when I pick up this paperback book, worn out by hours of use and handling that was not always respectful (translation obliges), I am overcome with a feeling of gratitude and wonder at the memory of what has been for me a long familiarity with the scientific and metaphysical thought of a man who was both a great Christian and an exceptional scholar, in whom converged a knowledge that commands admiration and a deeply held Catholic conviction, passionately expressed. Indeed, not content with this criticism of Teilhard (and in particular his obsession with evolution), and due to circumstances that are of little importance here, having read almost all of his writings, I discovered in him new ideas relating to scientific questions that remain unresolved today, questions that physicists have been grappling with for almost a century—and this because their science has deliberately distanced itself from a global vision of the universe: that of what is known as "traditional" science. It seems that Wolfgang Smith is not only a great physicist, but also the only one in whom metaphysics and physics can be read in parallel or constantly intersect. Of course, this does not (in general) overshadow the scientific authors who, over the centuries, have contributed concrete and still valid knowledge about the world we live in. But in spending time with W.S. and through his conception of the integral universe, which he describes as "tripartite" (composed of three worlds—a vision very close to Hindu doctrine), we cannot help but become aware of our permanent anchoring in an immeasurable Reality, inaccessible to our senses, yet present in every one of our cells. One need only read his exposition on visual perception to grasp the astonishing complexity of this phenomenon, which involves levels of reality that classical descriptions completely ignore. We become aware that the immanence of a transcendent reality concerns not only our spiritual life, but also the humble deeds and actions of our daily lives. And this, with all due respect to certain theologians who seem to think that once our planet has been launched into outer space, the great Architect no longer cared about it. Yet everything comes from Above, from "aeviternity," the eternal "now" from which time and space as we experience them derive—and the most imponderable mechanisms of our earthly life are inscribed in the three levels of what W.S. calls an "irreducible Wholeness." The universe cannot be fragmented. The interest of Wolfgang Smith's solidly argued and absolutely fascinating demonstrations derives from the fact that they are based not only on a thorough knowledge of contemporary physics and mathematics, but also on two thousand years of metaphysical conceptions, which he shows us to be compatible with the most recent discoveries. They reveal the vanity of a so-called "knowledge" which, through excessive rationality and contempt for the supposed "infantilism" of ancient sciences, has locked us into a world that is, in Whitehead's words, "a jumble of particles moving endlessly, without meaning."

Marie-José Jolivet
France, November 3, 2024

❋ ❋ ❋

Almost twenty years ago now, after Pierre-Marie Sigaud had the brilliant idea of publishing *Sagesse de la cosmologie ancienne: Les cosmologies traditionnelles face à la science contemporaine*,[1] I had the chance to meet Wolfgang in person for the first time, as my professional activities took me to California several times a year.

This complemented regular but infrequent telephone conversations, as Wolfgang did not have email at the time, despite my insistence—a situation that, fortunately, would be rectified later.

Very quickly, needless to say, a deep friendship, even a bond, brought us together, and Wolfgang never failed to discuss his projects with me as they progressed. What an honor! What prestige! Rest assured, Wolfgang rarely, if ever, took into account the advice he sought from me. He stayed the course he had set for himself, followed his path, and pursued his mission. He had to fulfill his destiny, as he put it to a Brazilian friend.

This destiny was twofold: to defend the genius of Christianity against Orientalism and perennialism, and to defend science against all scientism. The common thread? Metaphysics!

[1] Paris: L'Harmattan, *Théôria* series, 2008. Translation of *The Wisdom of Ancient Cosmology: Contemporary Science in Light of Tradition* (Oakton, VA: Foundation for Traditional Studies, 2003).

This was also the common ground between Jean Borella, a metaphysician with a degree in physics, and Wolfgang, a physicist with a solid understanding of metaphysics. This "complementarity" would give rise, a few years later, to *Physics and Metaphysics*,[2] bringing Jean and Wolfgang together.

We dare to think that our friendship with Wolfgang transcends the limits of space and time.

Bruno Bérard
France, November 3, 2024

❋ ❋ ❋

Professor Smith and his writings were an integral part of my training as a scientist with a religious mind. I understood how the findings of modern science could be contextualized by drawing inspiration from the philosophical schools of many great figures, and how the reception of modern science could be criticized from many points of view.

In doing so, he enabled me to fully appreciate my career in physics. I can only imagine that he and his work have done the same for generations of students besides myself. If the Lord said that he would not forget the reward of one who gave his servants as little as a cup of water, he certainly cannot forget to bless one who has given so much.

May God bless him.

Olumakinde Ogunnaike
USA, January 18, 2025

❋ ❋ ❋

One of the great honors of my life, has been the privilege of knowing Professor Wolfgang Smith. My voyage into his extraordinary insights started after I encountered a podcast

[2] Paris: L'Harmattan, 2018; simultaneous publication in English: *Rediscovering the Integral Cosmos, Physics, Metaphysics, and Vertical Causality* (Brooklyn, NY: Angelico Press, 2018).

discussing his remarkable work, featuring the late Richard DeLano. In listening I was instantly captivated—not merely by Smith's masterful resolution of The Quantum Measurement Problem, but also by the sheer breadth of topics that Smith had addressed throughout his career. In today's academic climate it is nigh-on impossible to find an individual who has made significant contributions to fields as diverse as mathematics, physics, philosophy and theology. However, when it comes to Wolfgang Smith this is precisely what he achieved!

At just fourteen years of age, Smith read A. N. Whitehead's book *Science and The Modern World*, a book that profoundly influenced him and laid the foundations for his future work in philosophy. After graduating from Cornell at age eighteen, Smith enjoyed an illustrious career as a mathematician and physicist—proving theorems in differential geometry; yet, it will undoubtedly be Smith's work as a philosopher of science that he will be most remembered for. Through a groundbreaking series of books, beginning with *Cosmos and Transcendence*, Smith demonstrated how the ancient philosophical traditions not only align with physics but also resolve some of its most fundamental quandaries. In time, I believe Wolfgang Smith will be celebrated as the visionary who forever transformed mankind's understanding of the cosmos. May the writings in this book honor his legacy.

John Taylor
United Kingdom, April 24, 2024

❋ ❋ ❋

Some people's lifelong pursuits—the questions they chase, the ideas they value, the thoughts they find meaningful and true, and their spiritual journeys—bear striking similarities. I believe this is why I have come to love and admire Professor Wolfgang Smith, whom I first learned of through his books, and later had the privilege, even if from afar, to engage with in both written and spoken exchanges.

For those who are not content with simply marveling at the technological and worldly wonders of modern science, but rather are drawn to its intellectual and spiritual implications—and who, at the same time, feel compelled to compare the dizzying developments of recent centuries with the historical, cultural, philosophical, and religious traditions of the civilization into which they were born, to construct a coherent worldview—the situation is likely the same.

Smith was one of the foundational pillars whose thoughts and explanations brought peace to my soul regarding the religion-science conflict that I had felt since childhood and struggled to resolve satisfactorily. Specifically, he grounded the fundamental responses of Traditionalists such as Guénon, Schuon, and Nasr to questions like: What is modern science, what does it mean, what does it imply, and how trustworthy is it?—especially within the context of physics. In my view, what Smith accomplished in physics through his works *The Quantum Enigma* and *The Wisdom of Ancient Cosmology* parallels what Guénon did for mathematics in his *The Metaphysical Principles of the Infinitesimal Calculus*. I firmly believe that the value of this exceptional contribution of his, in the sight of God, is beyond measure.

I consider it a duty to once again express, on this occasion, my gratitude and heartfelt appreciation for him, and I pray that God reunites us in the hereafter among those with whom He is well pleased.

Ali Sebetci
United States, April 16, 2025

* * *

Wolfgang Smith has been important in my spiritual journey. Thinkers in the tradition of *sophia perennis* —notably Frithjof Schuon and René Guénon—made it possible for me to see through the errors of the age (and my own), clearing the ground for me to come to the Catholic faith, but he was the first specifically Catholic author I came to know who wrote in this tradition. *Cosmos and Transcendence: Breaking Through the Barrier of Scientistic Belief* developed that tradition further through its key distinction between science and scientism, which is to say: it untangled the confusion of hard science and false philosophy of science. With its confutation of Cartesianism and its exposure of the *a priori* rejection of causality outside of time implicit in evolutionism, it alerted me to how the prevalent contemporary epistemology saw the world only in its aspect of usefulness and not at all as showing (in Saint Paul's words) "the invisible things" of God. Together with the reflections of the French Catholic philosopher, Jean Borella, its thinking opened up for me a new way of beholding, one which is essentially contemplative. From this came my own book, *The Meaning of Blue: Recovering a Contemplative Spirit*, that was published by Angelico Press. This is the very press that in its early days, I was excited to discover, was precisely the source of further books by Wolfgang Smith. Although his work is now issued elsewhere, its predominance among the publisher's earliest titles was an auspicious seed of what was to come from this powerhouse of deep traditional thought.

Wolfgang Smith's works' special gift to the intellectual life of believers is the connections it makes. He was both a physicist and a metaphysician who was able to show the congruence of modern physics and traditional metaphysics, once spurious philosophy of science and prejudice against vertical causation are abandoned. This is evident in *The Quantum Enigma: Finding the Hidden Key* and is developed in the later work *Physics and Vertical Causation*, which contains the magisterial assertion that "nothing whatsoever can in fact exist without being 'vertically' caused." This of course is consonant with the teaching of Saint Thomas Aquinas that God is entirely present in each thing as its cause. The particular value of Wolfgang Smith's explanations is that one is no longer tempted to think of what theology teaches as unrelated to the understanding we receive through science. This is further exemplified in *The Wisdom of Ancient Cosmology: Contemporary Science in the Light of Tradition*, which highlights the implication of Bell's theorem that reality is non-local. Although Einstein himself did not come to terms with this, it is (Wolfgang Smith tells us) "precisely in the light of Einsteinian relativity that the trans-spatio-temporal nature of non-locality can be discerned." It follows that for physics as well as for faith, it makes perfect sense that a holy person (such as Padre Pio, for example) can be in more than one place at once. Most critical of all is the concomitant understanding of how Christ's redemptive work, which is eternally present, transcends the limitation of time. We already get a glimpse of this transcendence in Einstein's discovery that the closer anything gets to the speed of light, the more slowly time passes relative to what is not traveling at that speed. It is therefore entirely coherent to say that the Light of the Word Himself is "the same yesterday, today and forever." Difficulties in imagining how what happened in historical time has now an efficacious presence are dissolved in understanding that time and space are, as physics teaches, not absolute realities. Absolute reality is transcendent.

Wolfgang Smith's ability to make connections is also evident in *In Quest of Catholicity*. This is a book of extraordinary interest since it touches on nothing less than the intellectual credibility of the Catholic faith. Presenting this faith in a system narrowly contained in time and space can only have a provisional value. For the serious seeker, any truth claim has to be congruent with the most ancient wisdom of humankind as well as the insight of contemporary science. In this very personal dialogue with a priest whose overriding concern is the welfare of souls, Wolfgang Smith opens up for us the possibility of engaging with that wisdom wherever it might be found, as for example in the Hindu tradition, rather in the manner in which the early Church engaged with ancient

Greek thought. Such openness connects us with those who are looking not simply for the formulations of the past, but the saving truth which inspired them. His writing initiates a dialogue in which souls in our time can find it.

Somewhere Wolfgang Smith comments that although we hear of saints of our own days, it is seldom that they are placed before us as also being sages. God never deserts His Church, never refuses His light to humanity: nonetheless, our readiness to receive what He gives can be diminished not only by ill-will but also by an ideologically locked mind, whether that rigidity comes from scientism or from a perspective like that of the builders of Babel who could not communicate because they imagined their idiolect was the only tongue. We have cause to be deeply grateful to Wolfgang Smith for helping us to unlock our minds to holy wisdom and to open up the prospect of, once again as of old, sanctity and sagacity walking hand-in-hand.

Father Luke Bell
United Kingdom, August 30, 2025

❋ ❋ ❋

A long-awaited encounter between science and philosophy[3]

The issues addressed by scientists become all the more fascinating when they touch on philosophy. This is the case with Wolfgang Smith's beautiful book *Science, Scientism, and Religion*, which introduces us to a question that is undoubtedly unthinkable today: is our view of the world based on the "indisputable facts of science"? If we "reduce reality to physics or the measurable," this seems indisputable. "But the world cannot be reduced to quantity alone... which leads the human mind to wonder *who* defines what is 'indisputable'":

[3] See *Science, Scientisme, et Religion* (Paris: L'Harmattan, 2024), published in English as *The Vertical Ascent: From Particles to the Tripartite Cosmos and Beyond* (Philos-Sophia Initiative Foundation, 2021). Following citations refer to this text.

> While this view of the world is reputed to be based on "the indisputable facts of science," it is this view itself that ultimately determines what is or is not an indisputable fact. In the end, we are all metaphysicians, whether we admit it or not; it is not without irony that those who claim to despise metaphysics are generally determined to justify their own. (58–59)

Distinguishing between levels of reality in order to unite them

Starting from this philosophical questioning of our worldview, let us retain from our physicist and philosopher a fruitful idea expressed in the form of an aphorism: the measurable cannot measure itself. In other words, there is a fundamental dichotomy between physics—particularly quantum physics—and the world that Smith calls "corporeal," that cannot be violated. This raises a very pertinent question: if there is an observable world with its chain of horizontal causalities, the level of phenomena itself, what about a totality that precedes spatio-temporal fragmentation and ensures the coherence of micro and macro physical realities? What explains, in particular, what is known as the principle of non-separability examined by Niels Bohr and on which he disagreed with Albert Einstein?

> It is this concept of "fullness"—the notion of a totality whose overall behavior is richer than what would be obtained by the sum of its parts—that requires reflection... Authentic ontology requires such "fullness," and I believe that it causes, and in a sense explains, the phenomenon of non-locality—an apparently instantaneous action at a distance, occurring in the quantum world. (108)

In other words, considering wholeness requires an additional step on the part of the mind. If the principle that "a *whole* that cannot be reduced to 'the sum of its parts' cannot be fully described in quantitative terms" (108) is true, then another level of reality is needed: physics does not explain everything. Using the expression "*unbroken wholeness*," Smith argues for distinguishing between the physical system characterized by quantitative attributes and the observer, who is himself endowed with qualitative attributes. Thus, the encounter between these two distinct ontological planes—the physical realm itself and the observer, who is himself corporeal—is the result of "the actualization of a physical state that constitutes a transition from potentiality to actuality... This leads quite naturally to an ontological understanding of quantum mechanics" (111).

Let us welcome here a possible and desirable encounter between science and philosophy, with two distinct types of causality, with priority at the ontological level, the whole taking precedence over the parts, particularly in the phenomena of non-locality observed in quantum physics. Addressing the question of the totality of the world as it precedes spatio-temporal fragmentation, capable of ensuring the coherence of micro and macro physical realities, is a feat that requires a courageous effort of thought. In short, for Wolfgang Smith, what science cannot explain must leave open the idea of vertical causation, thanks to which quantum phenomena belong to a certain totality that escapes phenomenological measurement. How, indeed, can we explain the *decoherence* observed by scientists at the quantum level, which does not affect the macrocosmic domain? The answer: "The quantum world is made up not of particles, but of *potentiae,* halfway between being and non-being, to quote Heisenberg once again" (49). This term attempts to explain an ontological reality, and Wolfgang Smith uses the word "entanglement" to highlight this vertical intervention. The British philosopher and mathematician Alfred N. Whitehead used the term "ingression" to signify the participation of entities at a higher level of reality:

> It is by participating in eternal things that temporal things arise. The two series are mediated by something that combines the actualization of what is temporal with the

timelessness of what is potential. This final entity is the divine element of the world.[4]

Ontology sheds light on physics

Ontology sheds light on physics and its mysteries, even if it does not explain the phenomena observed. In any case, in a certain way, it provides the framework for their conditions of possibility.

Vertical interaction therefore brings about what was only potential, for example the position of the particle in relation to the wave function; hence the term *potentiae*, which is reminiscent of Aristotle's concept of act/potency. What is original about this approach is that it goes beyond physics to an ontological level. We are clearly dealing here with an interaction which, beyond the relationship between man and the nature he observes, introduces the metaphysical plane of being. He is led to this once again by the fact that the cosmos cannot be described exclusively by physics. Wolfgang Smith expresses this through the image of a circle divided into three domains: the aeviternal center where all totality resides, the circumference representing the corporeal world, and, between these two extremes, an intermediate ontological domain (145–46); this intermediate domain would then evoke, as in Thomistic metaphysics, the domain of being that receives itself from the eternal world and radiates into essences. To briefly recall the logic, being does not subsist by itself, but is the medium in which all things have their place, proceeding from Being and receiving their luminosity from it, which is another way of applying Thomas Aquinas's principle: Being accepts to be "finitized, to be 'received, limited, and circumscribed' in the essence of the finite creature," in the words of renowned Thomist David Schindler.

[4] *Process and Reality: An Essay in Cosmology* (NY: The Free Press, 1978), 350.

Here again, Alfred N. Whitehead refines and corroborates this vision, which is both physical and philosophical, when he completes the meaning of "ingression" as that which "designates the particular mode in which the potentiality of an *eternal* object is realized in a particular actual entity, contributing to the definite character of that actual entity."[5]

Thanks to Wolfgang Smith for allowing his readers interested in science to return to being. It is the possibility of reflecting on the surprising duality of contradictory phenomena in such a way that the qualitative thinking of being assumes realities that science cannot elucidate in its quantitative order.

P. Richard Escudier
France, September 14, 2025

[5] Ibid., 23.

Part One
Personal details

Chapter I

Significant Biographical Details

The information gathered here is succinct, drawn mainly from Wolfgang Smith's recently published memoirs[1] *as well as from private communications from his close friends. While more exhaustive biographies are invaluable for gaining a deeper understanding of an author's thoughts, the essential information presented here provides a good overview of the fields explored by Wolfgang Smith during his long career and, above all, the spirit that underpins his entire body of work.*

Towards a career in mathematics

Wolfgang Smith, born in Austria in 1930, emigrated to the United States with his family during his childhood, just after the outbreak of World War II. He was gifted with exceptional intelligence and insatiable intellectual curiosity, but above all, he was an *integral* thinker, with a rare dual expertise in the "hard sciences"—mathematics and theoretical physics—and metaphysics...

Starting with science, he quickly earned bachelor's degrees in mathematics, physics, and philosophy (Cornell University), a master's degree in theoretical physics (Purdue), and a doctorate in mathematics (Columbia). At the same time, he conducted aerodynamics experiments in industry at Bell Aircraft Corp., contributing to the development of procedures for the reentry of space shuttles into the atmosphere by manipulating the porosity of the materials used in the

[1] *Vedanta in Light of Christian Wisdom* (Philos-Sophia Initiative Foundation [hereafter PSIF], 2022).

hulls of the spacecraft in order to better dissipate intense heat. A brilliant idea!

However, his career was mainly academic until his retirement in 1992, as a professor of mathematics at MIT (*Massachusetts Institute of Technology*), UCLA (*University of California, Los Angeles*), and OSU (*Oregon State University*), marked by research in differential geometry and topology, the results of which were regularly published.[2]

It should be noted that mathematics was initially the only choice left to him in terms of academic discipline and "professional activity," but he never regretted it. "Only choice" because, at Cornell University, he was deeply disappointed by analytical philosophy. He sums up this disappointment with Max Black's (1908–1988) response[3] to his remark about a yogic state in which the separation between man and God is overcome: "I don't deny that this may be a great experience, but it can't have anything to do with the truth."[4] He experienced the same disappointment with physics when, this time at Purdue, he discovered during his master's degree in theoretical physics that physics itself had become *purely* analytical.

He repeatedly expressed his good fortune in having devised his own mathematical theory of submersions,[5] which allowed him to coexist with his colleagues in the academic world[6] without having to face controversy.

[2] By the renowned American Mathematical Society, both in its *Quarterly Bulletin,* which has been in existence since 1891, and in its *Transactions,* the monthly journal that has been in existence since 1900, and in journals such as the *Proceedings of the National Academy of Sciences,* the *American Journal of Mathematics,* the *Tohoku Mathematical Journal* of Tōhoku University in Japan, and the *Pacific Journal of Mathematics.*

[3] A philosopher of science and language, Black is considered a pillar of American analytical philosophy.

[4] *Vedanta in Light of Christian Wisdom,* 1–3.

[5] In differential topology, a submersion or submersive map between two differential manifolds is a differentiable map whose differential is surjective at every point.

[6] Samuel Bendeck Sotillos, private communication.

Metaphysics: from Vedanta to Christianity

With the aforementioned "yogic state," one leaves "modern science" behind for metaphysics. Long before his pilgrimage to India, he was deeply moved by reading the first poem in the collection *Gitanjali / Song Offering* (1912),[7] composed by Nobel Prize winner Rabindranath Tagore (1861–1941):

> Thou hast made me endless, such is thy pleasure.
> [...] At the immortal touch of thy hands my little heart loses its limits in joy and gives birth to utterance ineffable.

Of course, he didn't stop there! While studying, Wolfgang discovered Rāmakrishna and got in touch with the Rāmakrishna Mission founded by Swami Vivekananda (whom Rāmakrishna named as his successor). He was 18 and started going to Vedanta Society meetings in New York, under the guidance of Swami Nikhilananda (1895–1973). From then on, he began to "see the world through Vedic eyes,"[8] strengthened by the conviction that there is ultimate knowledge, "knowledge by which all things are known," as indicated in a passage from the *Vedas.*[9]

After earning his doctorate in 1960, he left MIT to spend seven "privileged" months among monks, the *sādhus,* notably Śrī Ānandamayī Mā (1896–1982), one of the great spiritual figures of the twentieth century, approaching them with "deep reverence" and receiving unforgettable "spiritual gifts" from them.[10] However, even at that time, Smith understood that viewing Christianity through the eyes of the Vedas introduced a bias into the analysis, tended to reduce Christianity

[7] In French, *L'Offrande lyrique,* NRF, 2014, translated from English by André Gide.

[8] *Vedanta...,* 2.

[9] A collection of texts revealed to *the Rishi* and divided into four works: *Rig-Veda* (c. 1500–1200 BC) and *Yajur-Veda, Sama-Veda,* then *Atharva-Veda* (1200–900).

[10] *Vedanta...,* 4, and Samuel Bendeck Sotillos, private communication.

to a "more or less exoteric" level, and, above all, led to an undue confusion between what he called the nirvanic option of Vedānta and Christian salvation. It was only halfway there.

It was his relatively late marriage to Thea,[11] a young woman who was "deeply Christian," that was one of the triggers that prompted Wolfgang Smith to take the other half of the journey. Married after she obtained her doctorate in mathematics at the university where Wolfgang Smith taught, Thea was an extraordinary woman (whom I had the pleasure of meeting), very talented, also a musician, dancer, and singer. She was the primary influence on Wolfgang's return "to his Christian roots." To this we must add the two key readings that he himself mentioned: a biography of St Catherine of Siena and St Augustine's *Homilies on the Gospel of St John*. We should also mention a visit to Padre Pio (1887–1968) in San Giovanni Rotondo, whom he considered one of the greatest modern saints of the Catholic Church, by whom he felt blessed, and the privilege of attending a public audience with Pope Pius XII (1876–1958) in Rome.

However, it was only some sixty years after his "initiatory" trip to India that Wolfgang would resolve this question between Vedanta and Christianity. He renounced the trap into which, he wrote, Henri Le Saux (1910–1973) and Frithjof Schuon (1907–1998) had fallen:[12] identifying *nirvana* and salvation purely and simply—and abandoned the two approaches of observing each religion through the eyes of the other.

Wolfgang draws the basis for such an approach—looking without a point of view, observing without a point of observation—from a metaphysical exegesis of Meister Eckhart, which he describes as "a masterstroke" and "Christian yoga."[13] From

[11] Dorothy Marguerite Smith, born in 1926 and died at the age of ninety on June 21, 2017.

[12] *Vedanta*..., chap. 2, "From Benedictine to *Sādhu:* the case of Henri Le Saux," and chap. 4, "Breaking the *ātma-māyā* Dichotomy."

[13] Ibid., 5, 8, 35ff.

then on, the Vedantic dichotomy *of ātma-māyā* dissipates in the Trinity and the Incarnation. Salvation, unlike *nirvana*, is the fruit of the mediation of the Incarnate Word.

With his deep understanding of Christianity, Wolfgang Smith was able to carefully critique the theistic evolutionism of Teilhard de Chardin (1988),[14] explain what Christian gnosis is (2009),[15] publish two years of correspondence with Father Malachi Martin (1921–1999) (2016),[16] and finally, to distinguish Christianity from perennialism (2022).[17]

From metaphysics to physics

Of course, there is more than just this religious metaphysics—ultimate though it may be—because Wolfgang developed a passion for physics at a very early age and, through Platonic and Neoplatonic philosophies, for the philosophy of science. Eight years before his retirement, he began his work with a first book on integral cosmology (1984).[18] Over the years, he has opposed the limitations to which contemporary science has subjected itself with a superior science, in which man is *ultimately* both the observer and the scientific instrument. As we read, man

> becomes, in a way, the "microscope" or "telescope" through which he is able to perceive the invisible expanses of the integral cosmos.[19]

[14] *Teilhardism and the New Religion. A Thorough Analysis of the Teachings of Pierre Teilhard* (Tan Books, 1988), reprinted as *Theistic Evolution: The Teilhardian Heresy* (Angelico Press, 2012, PSIF, 2023), translated by Marie-José Jolivet: *L'évolutionnisme théiste de Teilhard de Chardin* (L'Harmattan, *Théôria* collection, 2023).

[15] *Christian Gnosis: From Saint Paul to Meister Eckhart* (Angelico Press, 2009).

[16] *In Quest of Catholicity: Malachi Martin Responds to Wolfgang Smith* (Angelico Press, 2016, PSIF second edition, 2024).

[17] *Vedanta*... op. cit.

[18] *Cosmos & Transcendence: Breaking Through the Barrier of Scientistic Belief* (Sherwood Sugden, 1984; Angelico Press, 2008; PSIF, 2022).

[19] Website of the *Philos-Sophia Initiative Foundation*, https://philos-sophia.org /wolfgangsmith/.

And he gradually applied this metaphysics—a traditional ontological realism—to quantum physics, cosmology, visual perception, and evolutionism. To achieve this, it was necessary to explore and critique the philosophical foundations of contemporary science, which are at the root of scientific reductionism or scientism,[20] in order to offer an integrative vision of reality, reconciling scientific and spiritual perspectives. It should be noted that for Wolfgang Smith, it is never a question of "rejecting a single iota of legitimate science," but simply of purifying it of an ephemeral and undue episteme and disentangling it from reductive and castrating scientism... For him, it is a question of restoring science's freedom and nobility.

From 1999 to 2008, Wolfgang Smith wrote numerous articles (14) regularly published by *Sophia, The Journal of Traditional Studies* (Oakton/VA, USA), covering topics such as matter, cosmology, visual perception, and approaches related to science (myths, esotericism, Kabbalah, Gnosticism, etc.).[21] These articles, sometimes adapted, along with the text of a lecture published in *The Thomist* 63 (1999), were translated to form *Sagesse de la cosmologie ancienne* (*Wisdom of Ancient Cosmology*). *Les cosmologies traditionnelles face à la science contemporaine* (*Théôria,* L'Harmattan, 2008, with a preface by Jean Borella), later republished in the United States as *The Wisdom of Ancient Cosmology: Contemporary Science in Light of Tradition* (Foundation for Traditional Studies, 2009, preface included).

In the 2010s, it is worth noting *Science and Myth: With a Response to Stephen Hawking's* The Grand Design,[22] whose French subtitle ("From Physics to Science Fiction") speaks for

[20] For example, *Ancient Wisdom and Modern Misconceptions: A Critique of Contemporary Scientism* (Angelico Press, 2013; PSIF, 2023).

[21] Summaries of articles by Wolfgang Smith can be found at metafysikos.com in five languages.

[22] Angelico Press, 2012.

itself, and, co-written with Jean Borella, *Physics and Metaphysics*,[23] bringing together a metaphysician with a degree in physics and a physicist who became a metaphysician.

Despite the great pain caused by the loss of his wife, health problems (heart-related) and accidents leading to hospitalizations, Wolfgang saw through what he had to write. "When we are born, we have a destiny, a path we are supposed to follow; we are free to comply with it or not. My interest in physics and mathematics gave me what I was meant to do. Following this path, as God had entrusted to me, was more beautiful than anything one can imagine... my task was to "fulfill my destiny," he confided to his friend, Samuel Bendeck Sotillos.

He therefore created a foundation, the *Philos-Sophia Initiative Foundation* (PSIF), whose stated mission is nothing less than "to promote the restoration of authentic philosophy, based on the ontological resolution of the problem of quantum reality"... "*Ad majorem Dei gloriam*"; and to publish three final books:

- *Physics and Vertical Causation. The End of Quantum Reality* (Angelico Press, 2018; PSIF, 2023), which concludes, thirty years later, his early work from 1995 on the "quantum enigma" (cf. *The Quantum Enigma: Finding the Hidden Key* [Open Court Pub Co, 1995; Sophia Perennis, 2005; PSIF, 2023]).
- *The Vertical Ascent. From Particles to the Tripartite Cosmos and Beyond* (2021)
- *Physics: A Science in Quest of an Ontology* (2023).

Just before Christmas 2023, after a bad fall, Wolfgang Smith had to be admitted to palliative care in a specialized home in Camarillo, his hometown north of Los Angeles, from where he was born into Heaven on July 19, 2024. Having lived for

[23] L'Harmattan, *Métaphysique au quotidien*, 2018, introduction by Bruno Bérard, translated into English under the title *Rediscovering the Integral Cosmos: Physics, Metaphysics, and Vertical Causality* (Angelico Press, 2018).

decades a life of prayer and contemplation, and never failing to place his trust in God, he would have wanted to be remembered "as a believing Christian," he confided to a friend.[24]

For his tombstone, Wolfgang chose the following inscription *Deus Caritas Est* ("God is love"),[25] "because all things are encompassed in divine mercy," he said, insisting that *love* is not a mere sentimental notion, but a way of life."[26]

Let us add, in terms of posterity, the announcement just made by Professor Anthony McClaran, Vice-Chancellor of St Mary's University in Twickenham: the creation of the Wolfgang Smith Chair in Philosophy, with Dr. Sebastian Morello in charge. The aim will be to "explore the intersections between science, classical metaphysics, and the Christian intellectual tradition," particularly "in the key areas to which Smith himself contributed, with an emphasis on areas where fruitful work remains to be done."[27]

Bruno Bérard

[24] Samuel Bendeck Sotillos.

[25] See "God is love, and whoever remains in love remains in God and God in him" (1 Jn 4:16).

[26] See below, Samuel Bendeck Sotillos, "An Encounter with Wolfgang Smith."

[27] "New Wolfgang Smith Chair in Philosophy established at St. Mary's University," stmarys.ac.uk/news/2025.

Chapter II
Essential Bibliography

We do not include here the scientific articles published by Wolfgang since 1959 in mathematical journals, because they are easily available online and, above all, because it is what goes beyond science in the strict sense that interests us here. As we shall see, there is already enough to deal with in those articles and books published in his remaining body of work.

For the record, Wolfgang's mathematical research in differential geometry and algebraic and differential topology has been published by the renowned *American Mathematical Society*, both in its quarterly *Bulletin,* which has been in existence since 1891, and in its *Transactions*, the monthly journal that has been in existence since 1900. Other journals include the *Proceedings of the National Academy of Sciences, the American Journal of Mathematics,* the *Tohoku Mathematical Journal* of Tohoku University in Japan, and the *Pacific Journal of Mathematics.* With publication dates ranging from 1959 to 1980, it can be said that they accompanied Professor Smith's mathematical work almost until his retirement in 1992.

The reason these publications stopped in 1980 was because Wolfgang turned his attention to cosmology and religion and, in particular, to defending both in the face of an invasive and deleterious scientism. He published two books in the 1980s:

- *Cosmos and Transcendence: Breaking Through the Barrier of Scientistic Belief* (Sherwood Sugden & Company, 1984), a French translation of which is due to be published in 2026,[1] and

[1] French trans. Marie-José Jolivet, coll. *Théôria*, L'Harmattan.

• *Teilhardism and the New Religion* (Tan Books & Publishers, 1988), which was updated and republished under the title *Theistic Evolution: The Teilhardian Heresy* (Angelico Press, 2012) and has just been translated into French: "L'évolutionnisme théiste de Teilhard de Chardin" (coll. *Théôria,* L'Harmattan, 2023).

In the 1990s, at the beginning of his retirement, he wrote his famous *The Quantum Enigma. Finding the Hidden Key,* published by Sherwood Sugden & Company in 1995, which, with a preface by Seyyed Hossein Nasr, was republished by Sophia Perennis in 2005.

The 2000s were extremely productive. They include his feature articles, published mainly by *Sophia, The Journal of Traditional Studies,* as well as two books that focus on the fight against scientism and the understanding of Christianity:

• *The Wisdom of Ancient Cosmology: Contemporary Science in Light of Tradition* (Suhail Academy Lahore, then Foundation for Traditional Studies (2003, 2009), republished under the title *Ancient Wisdom and Modern Misconceptions: A Critique of Contemporary Scientism* (Angelico Press, 2004)

• *Christian Gnosis: From Saint Paul* to *Meister Eckhart* (Angelico Press, 2008).

Wolfgang's articles typically fall into two categories: contemporary science revisited in the light of traditional wisdom (science) and more specifically theological subjects (theology), to which we can add a few articles about specific people or books (*miscellaneous*):

Science

"Celestial Corporeality"	*Sophia* v.5	1999
"The Extrapolated Universe"	*Sophia* v.6	2000
"Eddington and the Primacy of the Corporeal"	*Sophia* v.6	2000
"Science and Myth: the Hidden Connection"	*Sophia* v.7	2001

"*Sophia Perennis* and modern science"	Chapter 12 in *The Philosophy of Seyyed Hossein Nasr*	2001
"The Pitfall of Astrophysical Cosmology"	*Sophia* v.7	2001
"Esoterism and Cosmology: From Ptolemy to Dante and Cusanus"	*Sophia* v.8	2002
"Modern Science and Guénonian Critique"	*Sophia* v.9	2003
"The Enigma of Visual Perception"	*Sophia* v.10	2004
"Neurons and Mind"	*Sophia* v.10	2004

Theology

"Cosmology in the Face of Gnosis"	*Sophia* v.12	2006
"Transcending the *creatio* ex nihilo: The Kabbalistic Exegesis"	*Sophia* v.13, 1	2007
"The Wisdom of Christian Kabbalah"	*Sophia* v.13, 2	2008

Miscellaneous

"From Schrödinger's Cat to Thomist Ontology"	*The Thomist* 63	1999
"Review of Journeys East, by Harry Oldmeadow"	*Sophia* v.11	2005
"Rama P. Coomaraswamy, 1929–2006: In Memoriam"	*Sophia* v.12	2006

The 2000s also saw a first publication in French, in 2008: *Sagesse de la cosmologie ancienne: les cosmologies traditionnelles face à la science contemporaine* (coll. *Théôria*, Paris: L'Harmattan), with a preface by Jean Borella.

2008 was indeed an important year. It was then that Pierre-Marie Sigaud, editor of the *Théôria* collection at L'Harmattan, came across Wolfgang's articles in *Sophia* and suggested that he compile fourteen articles to produce this first French edition of his work.

From 2012 onwards, there were ten years of important publications, which now complete Professor Smith's work:

- *Science & Myth: With a Response to Stephen Hawking's* The Grand Design, 2012; published the following year in French under the title *Réponse à Stephen Hawking. De la physique à la science-fiction* (2013);[2]
- *In Quest of Catholicity: Malachi Martin Responds to Wolfgang Smith* (Angelico Press, 2016);
- *Physics and Metaphysics*,[3] and simultaneously in the United States, *Rediscovering the Integral Cosmos; Physics, Metaphysics & Vertical Causality* (Angelico Press, 2018);
- *Physics and Vertical Causation: The End of Quantum Reality* (Angelico Press, 2019);
- *The Vertical Ascent: From Particles to the Tripartite Cosmos and Beyond* (Angelico Press, 2021);[4]
- *Vedanta in Light of Christian Wisdom* (PSIF, 2022);
- *Physics: A Science in Quest of an Ontology* (PSIF, 2023).

The latter two books are published by the foundation created by Wolfgang Smith with the help of his friend Rick DeLano;[5] the foundation has had a website since 2018: https://philos-sophia.org/.

Since the end of 2023—Wolfgang was delighted about this shortly before Christmas—his foundation has reissued all of his published titles with a standardized design. It should be

[2] Coll. *Métaphysique au quotidien*, with a preface by Jean-Jacques Flammang and translation by Ghislain Chetan.

[3] Ibid., translation by Ghislain Chetan, preface by Bruno Bérard (L'Harmattan, 2018).

[4] French translation by Marie-José Jolivet, coll. *Théôria* (L'Harmattan, 2024).

[5] Rick DeLano, now deceased, was a screenwriter and producer, known for the controversial film *The Principle*, 2014, and *The End of Quantum Reality*, 2020.

noted that the latest title is now completed by a third part consisting of the two important chapters mentioned earlier: "Platonist Cosmology: The Long-Lost Key" and "On Being and Consciousness," Wolfgang's very last unpublished works.

What was the purpose of this foundation?

The goal, as Wolfgang saw it, was to "have an impact on our civilization," no less, *at least* in the scientific realm, by helping us to "rediscover the "vertical" aspect of cosmic reality," with a view to "a global vision of the world that integrates the fundamental discoveries of physics into the traditional wisdom of humanity."

However, Wolfgang finally dissolved his foundation in the third quarter of 2023, content in the knowledge that his books would remain available.

Part Two
Physics and Metaphysics
Materia

Chapter III

Metaphysics into Physics

Top Down

One of the proofs of God's existence given to us by our good saint Thomas Aquinas is the argument from essential causal chains. This proof has deep implications for physics.

The classic example is a man using a stick to move a stone. The stone's movement is caused by the stick. The stick's movement is caused by the hand and arm. But inside the hand and arm are muscles, and these are moved by individual cells. The cells are "moved," i.e., changed, by the chemicals in the cells. The chemicals in the cells are changed or moved by the protons, neutrons, and electrons in them. These particles are themselves moved by quarks. The quarks are moved, some say, by strings. The strings are moved by possibly something underneath them.

And so on. But not and-so-on forever.

This *simultaneous* chain of essential causes has to bottom out somewhere. It cannot proceed to infinity, or no movement would ever get started. There has to be a base *first cause* in this chain. Without (again) going into the details, it turns out this first cause has to be the same first cause in *all* change. And, as the saint says, we call this first cause God.

Hold that in mind and let's next recall a version of an EPR experiment. Two entangled photons are released, one heading north and the other south. According to quantum theory, neither of these photons has a single state of polarization, but each photon has all possible polarizations, represented by its wave function. Eventually, the northward bound photon is interacted with, or "measured," which causes its wave function to "collapse" to a definite polarization. This interaction "causes"—the scare quotes are justified—the southward bound photon to take the opposite polarization *instantly*, even if the photons are so far apart that no signal could possibly communicate between them in time.

This fact about the world bothers many, because it doesn't appear there is any theory involving ordinary essential causes that can explain it. Bell even proved no such essential cause can exist, not if one embraced locality. And locality is definitely out: there's just no way for the photons to talk to each other.

Not if you heed exclusively to what Wolfgang Smith calls "horizontal causality," which is the type of causation ordinarily encountered in physics—the science, that is. However, if you allow "vertical causality," a metaphysical concept like, say, ordinary essential causality, the problem of how those photons can talk to each other goes away.

Smith does not mention the stick-moving-stone type of "horizontal" causal chains in *Physics & Vertical Causation: The End of Quantum Reality*, but these chains, which are in every movement or change everywhere and everwhen, point to a simple "vertical" cause that is omnipresent, like we met above, a cause not in space or time, a cause not limited by locality, because, as it were, all points and all times are available to this vertical cause always.

Physicists can't see this cause in their metaphysics, which excludes the spiritual in what Smith calls the corporeal world. If in your science all you have is a yardstick, then all that can exist is length. All other properties vanish, or rather are invisible. It would be an obvious fallacy to say that these other properties have no existence because they cannot be measured in your yardstick-science. But that is the central mistake Descartes built into physics when he separated the world into two domains, *res extensa* and *res cogitans*.

The first, *res extensa*, is the realm of the yardstick, the second, *res cogitans*, is the playground of the mind, where all those things that cannot be measured live. That is, the non-length properties are only in the mind, or are a projection of the mind, and are not in things themselves.

The opposite philosophy is realist and rejects these categories: it is the corporeal world of experience that is tangible, and it is the mathematical abstraction of the corporeal world's measurable or physical properties which is not. In other words, the modern science of physics rests on the Deadly Sin of Reification.

You can only get so far with horizontal causality. It nicely explains many measurable phenomenon of interest, but it can't explain everything. It can't, for instance, explain to us our consciousness. No yardstick or probe can ever exist to measure vertical causality, but we can still prove it is there. Prove it to yourself: that stone never would have moved by the stick if you didn't first will it. Will moves by vertical causality.

We also know the qualities of the corporeal world are there, because we experience them, even if they cannot be measured. Of course, we have met many examples of attempts to quantify the unquantifiable—how sanguine are you on a scale of –42,003.7 to e^{π}?—and we have seen how they all ultimately come to grief.

Smith's *Physics & Vertical Causation* is a culmination of his life's work, with each chapter referring to his more detailed earlier writings. He introduces hylomorphism, giving its outlines but not bogging the concept down with excessive detail, which would distract new readers. There are plenty of other texts that provide formal proofs of these concepts. Smith gives enough material to allow the reader a grip, and then moves on.

Bottom Up

Physics is stuck because of neglecting these matters. Ever since the so-called standard model settled into its comfortable form in the 1970s, not much new, or not much new that turned out true, has happened. This hasn't stunted the flow of activity or papers, of course, which is greater than ever. Academics must publish or perish. But there is an awareness something has gone wrong. That something, said Wolfgang Smith and others, was a faulty starting point.

Smith was one of the first in the movement to restore classical metaphysics to physics, as detailed above. Even though some strangely deny it, all physics must operate with a metaphysics, a philosophy of nature, even when this is denied or the metaphysics or philosophy is unspoken. The classical metaphysics which Smith championed is now re-blossoming, and this is in great proportion due to Smith.

He was humble to the point of shyness about his efforts. No, that's not the right word. "Fear and trembling" are better, and blessed with a superior provenance. Indeed, his writings evince a deep concern to reconcile his work with Catholic dogma, a set of beliefs consonant with his philosophy of nature. This is expressed best in the book *In Quest of Catholicity*, a collection of correspondence between Smith and Fr Malachi Martin. Smith had just given the 1998 Templeton Lecture on Christianity and the Natural Sciences, "From Schrodinger's Cat to Thomistic Ontology," which he had earlier worried his colleagues would not take well.

> The anticipated opposition—for which I had prepared—did not materialize... In the course of my lecture I sensed that the audience was attentive to every word; there seemed to be a certain "magic" in the air, an angelic presence I like to think. I felt that something was being transmitted and received. The questions which followed were for the most part intelligent and searching, never antagonistic.

The lecture followed 1995's *The Quantum Enigma*, Smith's first major work. Reading the lecture now (which you must), it seems wholly non-offensive. But you have to remember something about academics. Querulous is the nice word. The natural instinct of many is to snootily doubt everything you say, to dismiss it as useless, or to think they already thought of it. This is because, like every other human endeavor, academics is a competitive and often vicious sport. I do not say this is a bad thing: in fact, this fading confrontational approach is necessary to test theory against nature.

More Than Matter

"According to vulgar belief, there is color, the sweet and the bitter; but in reality, only atoms and the void."

So said Democritus, and so say many modern scientists. Smith was fond of this quotation. He argued the vulgar are right and the (materialist) scientists wrong: there is more than atoms and void. Or, rather, Smith said Democritus's critics were right, but only up to a point. That point is that where metaphysics necessarily takes over from physics. Smith realized that the solution to the many enigmas of quantum mechanics could not to be derived from an atomistic metaphysics. He urged, right up until the end, a return to the beginning, which is to say to an Aristotelian and Thomistic metaphysics, which Heisenberg himself first suggested.

There are two main curiosities about discrete-movement (i.e., quantum) physics. The first is the peculiar behavior of certain small things when they are (seemingly) separated from big things. They can at times appear to be many places at once. Only when they are interacted with by measurement do they take measured values. Their "wave function collapses," some say. This, following Heisenberg, Smith realized, as many now agree, could be explained by grasping that being can possess degrees of actuality and potentiality, Aristotle's solution to certain classical paradoxes. It's not that these tiny entities are now waves, and now particles. It's that "quantum" fuzz is *part* of their being. Quantum objects have much less actuality, like large objects have, and much more potentiality. An interaction, such as measurement, removes potentiality and forces an actuality.

Smith's resolution to quantum mechanical measurement is much easier, relatively speaking (he punned). This is to return to the beginning, to Aristotle and Heisenberg, to the idea that substances are composed of both potentiality and actuality. Even you, dear reader, have the potential to be somewhere other than where you are now, and it takes something actual to get you there, something actual to actuate the potentiality in you. This potentially is part of you, in a very real sense.

The quantum world is unfamiliar because it is composed of objects with much more potentiality than actuality. There is no wavefunction "collapse," but a potentiality becoming an actuality by something actual—that interaction or measurement. We spent a lot of time with this elsewhere, and there is no reason to repeat it here.

Smith's big idea, and one we share, is the return of the spiritual to scientific discourse. We all know it's there. "The first point to be made is that this reduction of the animate to the inanimate—of the living to the merely 'complex'—so far from being based upon scientific fact, is actually a groundless assumption, which gains strength from the fact that it is beyond our means to grasp whatever it may be that distinguishes the two."

One moment your crazy uncle is raving about BLM, the next moment he is not. His soul has departed. Living beings are more than their constituent parts working together, just like water is more than just hydrogen plus oxygen. "This soul-generated [vertical causation] constitutes in fact the life force or *élan vital* of the organism, which both 'produces' its body or 'corporeal sheath' and renders it *animate*." Any science that does not acknowledge this, and many other like things, is incomplete.

It is probably clear by now that vertical causality, unmeasurable in the traditional sense, is "*what gives rise to horizontal causality*." Look: you knew this was true before you came here today. All the gloried "laws" of physics and the nature of the world *had to come from something*. They could not make themselves. And we all know what this something is. Every attempt by scientists to explain this uncomfortable truth away has been an embarrassment. And always will be.

Enigmas

One can read Smith's *Quantum Enigma* to get detailed understandings of things, but for strictly introductory books, see Ed Feser's *The Last Superstition* and *Aristotle's Revenge*, as these assume little background and outline the philosophy more broadly, and not just for quantum mechanics. But if you already know physics, get Smith's. Indeed, get that book into the hands of working physicists, if you can.

The second mystery, and the one to which Smith devoted much of his efforts, is to explain how two "entangled" entities can communicate faster than the speed of light. The solution certainly isn't "many worlds," which Smith dismissed (as do I). The answer, with reference to Bell's theorem, is (again) that reality is more than atoms, or strings or whatever, bumping into one another. The visible cosmos is only the surface of reality. Underlying it is an Intermediate Realm, which itself is connected to the Prime Mover. This philosophy accords with all known observations, whereas the Democritian ideas do not.

A favorite image of Smith's was a circle, the circumference of which is, well, us. And bugs, rocks, stars, and all the rest. The interior is an intermediate realm that communicates with the Center. This is an apt metaphor. Think of two particles on the circumference, separated by a far distance, much too far for a signal to pass between them in good time. Attached to each particle is a line that plunges to the depth of the interior. The two lines are connected. It is here that *vertical* causation, said Smith, instantaneously interacts with the particles. There is no light speed barrier in the intermediate realm. Physicists say that interactions like this must mean reality is "non-local." Smith carried out the implications of that thought. Vertical causation, which is the result, is different than ordinary *horizontal* causation, which takes place only on the circumference and in time.

For those who have some background in physics, this quotation from *Quantum Enigma* might help:

> [I]t is nonetheless a mistake to speak of "chance" in reference to the microworld. As I have pointed out before, the collapse of the state vector—which singles out one particular eigenstate from an ensemble of eigenstates—is not actually comparable to the toss of a die; for whereas the latter constitutes a temporal process, indeterminate though it may be, the collapse of a state vector cannot thus be conceived. Let it be said apodictically that state vector collapse is not the result of a temporal process, be it deterministic, random, or stochastic. A higher order of causality enters the picture, which needs to be distinguished categorically from temporal causality in any of its modes; the so-called "collapse," it turns out, can be attributed no more to chance than to determinism, but actually entails a kind of causality which, strange to say, is "not of this world." (110)

The indeterminism of a die toss comes because we do not know all the causes and conditions of the toss. But when we take careful note of them, that indeterminism becomes determinism, as is easy to show. For instance, I with my father built a coin-flip machine that controls all important causes and conditions of a flip and makes the coin come up heads every time. But there is no knowing all the causes and conditions of certain quantum mechanical entities; this has been proved, too. There have been many theories about this, but, I submit, only the account of classic metaphysics, as Smith insists, holds up.

Alas, the concepts Smith explicates, important as they are, will not appeal to many working scientists, because they do not come *obviously* equipped with formula that can be manipulated. We have only begun to work practical ideas out. But that doesn't mean that they can't bear research fruit. R.E. Kastner and others have a recent Arxiv paper "Taking Heisenberg's Potentia Seriously." They demonstrate how "actual events can instantaneously and acausally affect what is next possible," which might bear fruit in showing the Intermediate realm works. If the reader has a solid physics background, get Robert Koons's *Is St Thomas's Aristotelian Philosophy of Nature Obsolete?* I'll let you guess the answer.

There is, of course, much more to all this. We have not even begun to sketch out what he defined to be corporeal objects, i.e., those of our ordinary perception, and physical objects, which are those parts of the corporeal that are amenable to quantification. A great fallacy of our age is to assume these are the same, that the unquantifiable can be quantified (by questionnaires, even!), or that all that is not quantifiable is unworthy of study. There is also Smith's concept of irreducible wholeness, which is related to hylomorphism, the idea that all objects are composed of matter and form.

Think Of That

There are many proofs our minds are not machines. The simplest is that machines have no self-awareness. We do. Some attempt an escape by saying our awareness is "illusion"—not grasping that it takes self-awareness to be able to have illusions.

One that will be new to some is that visual perception requires a "transcendence of *temporal bounds*." We do not see movement like a camera, frame by frame, but "*all at once*." *See* is doing double duty here. Parts of the body process the images, if you like, but it still takes a mind to *see* the images, and understand them for what they are.

There is a neat explication of Gödel's theorem, which again shows that we need a mind to see, and thus we are more than "collections of particles." I quote Smith from *Physics and Vertical Causation* in full because it is familiar territory for us:

> By virtue of Part I [showing that proofs in an axiomatized math can be numbered], we may assume that there exists a function P(m,n), defined for all natural numbers m and n, such that, for every m, P(m,n) is a propositional function of n (an algebraic statement depending on n, which may be true or false) and a function Π(k) which orders all mathematical proofs in the given axiom system.
>
> We now define the following propositional function: "There exists no k such that Π(k) proves P(w,w)." Since our enumeration P(m,n) of arithmetical propositions is complete, there must exist a natural number s such that P(s,n) is the aforesaid function. Now consider the proposition P(s,s): the first thing to note is that this proposition is unprovable (since our construction entails that "there exists no k such that Π(k) proves P(s,s)"; and the second is that P(s,s) is true: for indeed there exists no k such that Π(k) proves P(s,s).

Now if you can *see* that—comprehend it, grasp it, own it—you can also see that no computer ever can. It involves *infinities of thought*, which is impossible for any computer, which would be set chugging along never to realize the answer. Never as in *never.*

I like to say Gödel did it the hard way. We already knew there are true but unprovable statements in any axiom system. The axioms themselves. These are propositions everybody believes, but for which there is no proof. Not in the ordinary sense of that word. Again, we need to touch an infinite mind to know what cannot be proved by ordinary methods.

That's Heavy

Hold onto your black holes. Smith really does put some difficult questions to Einstein. "Either Einstein is right and the equations of classical mechanics need indeed to be revised, or the equations of classical mechanics are correct as they stand, and it is actually his 'relativistic' mechanics that prove to be false."

Here, dear reader, I am on less stable ground, not knowing as much about relativity as quantum mechanics, so I am not in a position to intelligently critique Smith. Einstein rejected the idea of absolute rest: Smith does not. Smith argues that Einstein assumed too much: given that the Michelson-Morley experiment did not find an orbital velocity of the earth, yet Einstein still assumed it was there, and so forth.

But it didn't stop with M&M. Others have since at least claimed to measure orbital velocity by, for instance, looking at Doppler shifts in starlight as (it is presumed) the earth moves. However, it is a possibility that the earth sits still and stars move. The measurements would be the same.

> What I wish to emphasize is that Einsteinian relativity is actually predicated upon the assumption that *there can be no such thing as an immobile reference frame, a K_0 "at rest"*: it is this denial that leads quite naturally to at least the special theory of relativity. But given that there exists not a shred of empirical evidence in support of that denial, one sees that Einsteinian physics cannot but be based ultimately on ideological grounds.

Well, what about that "not a shred"? Empirical verification of relativity must involve very sensitive measurement, given the effects are almost completely negligible for most things. "Misconceptions aside, the question whether special relativity has passed empirical muster proves an incurably technical issue. And no wonder, if at a speed of 1000 km/hr one needs to differentiate between 1 and 0.99999945!"

From here, Smith provides details of experiments (starting with the 1913 Sagnac experiment) showing anisotropy in light speeds. "Ruyong Wang, et al. conducted an experiment in 2003, in which the 'Sagnac effect is also obtained on a two-way linear path, by reversing a light beam sent out on a straight line on a moving platform and measuring the difference in return time.' What the Wang experiment indicates is that the speed of light is *not* in fact *c* in every inertial reference frame, as Einsteinian physics demands."

Science Is Not The Answer

Smith's other great passion, and perhaps yours, too, dear reader, was scientism. He didn't like it. We don't either. The idea that science has the last word on everything was to Smith, and to us, absurd. Science cannot even explain its own premises. This area is so large, and so dear to our hearts, that we will in future articles spell out Smith's objections. For now, one quotation from *Ancient Wisdom and Modern Misconceptions: A Critique of Contemporary Scientism* on the consequences of scientism:

> It is obvious to all that our outer lifestyles are being drastically altered as a direct consequence of the scientific advance. What we generally fail to realize, on the other hand, is that the impact of this same development on our inner lives—yes, on the condition of our soul—is no less pronounced. To begin with, the mechanization of our work environment, the phenomenon of urban sprawl, the rising congestion and perpetual noise, the proliferation of concrete, steel, and plastic, the loss of contact with Nature and with natural things, the invasion of our homes by the mass media—all this in itself is bound to have its effect on our mental and emotional condition. Add to this the uprooting of people from their ancestral environment, and unprecedented mobility which shuffles populations like a deck of cards!... For it must not be forgotten that people too have to be standardized, like interchangeable parts of a machine, so that the wheels of the mechanized civilization may run smoothly and efficiently.

Guided, he didn't have to say, by Experts under the thrall of utilitarianism, another gift of the Enlightenment. Of that, we know too well.

His books on this are *Cosmos and Transcendence: Breaking Through the Barrier of Scientistic Belief*, from which that passage was taken, *Ancient Wisdom and Modern Misconceptions*, and *Christian Gnosis: From Saint Paul to Meister Eckhart*. The last, and *Vedanta in Light of Christian Wisdom*, might be more properly classed as theo-philosophical.

I only "met" Wolfgang a few times, and only over Zoom and by email, where he very patiently and kindly corrected some of my misunderstandings of his work. I stress kindly. This attitude was ever-present, and obvious, in his writings and personal interactions. His correspondence with Fr Martin is saturated in humility. I look forward to science (re-) discovering his work.

William M. Briggs

Chapter IV

Wolfgang Smith's Philosophy of Physics

A Personal Reflection

According to Wolfgang Smith the cosmos we inhabit is composed of two essential realms.

The first realm, the *corporeal universe*, is the familiar world that greets us daily in our five senses of taste, touch, hearing, smell and sight. Importantly, Professor Smith argues that the qualities presented in our senses are authentic parts of external objects.[1] In other words, the sky up above is truly blue and the grass beneath our feet is genuinely green. Although this idea may seem naïve—something that only a "simpleton" would ever believe—Wolfgang Smith's notion of the corporeal world puts a mirror to the face of the dominant intellectual paradigm. A paradigm which insists on the bifurcation of reality into the perceptible and the imperceptible at every turn. Commencing with René Descartes and crystalizing with John Locke conventional wisdom now has it that sensory qualities are subjective projections rather than objective properties.[2] On this view, grass is not intrinsically green—since its greenness is merely a mental construct, superimposed onto whatever remains of the grass in an otherwise prosaic and colorless world. The bland remnants onto which sensory qualities are impressed, Locke and Descartes assume are the measurable attributes of external objects. What exists objectively, independently of the mind that is, are quantifiable properties such as length, width, and height. In the framework of modern

[1] *Physics: a Science in Quest of an Ontology* (hereafter PSQ), 8–9.
[2] Ibid., 3–4.

physics, these measurables also extend to highly augmented physical quantities such as those from the fields of thermodynamics, electromagnetism, and quantum theory.[3]

Naturally, this brings us to the second key realm from Smith's cosmos, the *physical universe.* In *The Quantum Enigma* (TQE), Smith defines the physical universe as the Cartesian domain of quantification—or, more precisely, the portion of reality accessible and intelligible to a physicist.[4] Typically, the *physical universe* maintains a one-to-one mapping with the corporeal world. As a result, Smith's ontology asserts that particular corporeal objects—such as apples in all their sensorial glory—correspond to subcorporeal objects, which represent their conceptualization in terms of physics.[5] To invoke Dr Smith's standard notation, for any corporeal object X, there exists a corresponding subcorporeal object SX, which is X as conceived by the physicist.[6]

However, this symbiosis between the corporeal and physical realms is not absolute in Smith's metaphysics. For hidden beneath the subcorporeal realm there lies an even deeper layer of physical reality: the transcorporeal domain. According to Smith, the transcorporeal domain encompasses those entities of physics that lack a direct origination from within an SX.[7] In his final book Smith associates these "transcorporeal objects" with quantum mechanical phenomena. A crucial point which we will return to in a short while.

To illuminate this further, Smith also introduces a concept that has since become foundational to his thought; that of irreducible wholeness (IW).

It will come as no surprise, at least to some, that Wolfgang Smith's philosophy of physics is deeply Thomistic and Platonist.

[3] Ibid., 4.
[4] *The Quantum Enigma* (hereafter TQE), 29.
[5] Ibid., 34.
[6] Ibid.
[7] Ibid., 28.

In fact, what individuates objects in the corporeal world are their Thomistic substantial forms—the ontological principles that choreograph matter (potentiality) into specific entities.[8] Put more poetically, a substantial form is that principle of actuality which makes, for instance, a soaring eagle different from a gust of wind, or a human being more than just flesh and bone, or a rational animal capable of love. The causality that substantial forms exercise Smith also states is fundamentally different from the causality found within physics.[9] Whereas in physics causality operates in time and in space by a sequence of events, a substantial form exercises a causality outside of time and space with no physical succession whatsoever. The first kind of causality Smith terms *horizontal causation* (HC) and the second kind *vertical causation* (VC).[10] In TQE, VC played a crucial role in shaping Wolfgang Smith's ontology of science. However, it wasn't until his last book that Smith provided a definitive account of VC's origin.

The account in question is rooted in none other than Smith's later conception of *Irreducible Wholeness* (IW). As the name clearly connotes, IW describes a wholeness which transcends a mere sum of spatiotemporal parts.[11] Now, the idea that the whole is greater than the sum of its parts has long been recognized by those with even a modicum of insight. However, throughout human history, this recognition has invariably been tethered to something else. For Aristotle and Aquinas, IW was tethered to substantial form, for Plato, to transcendent Forms outside the cave. However, unlike his venerable predecessors, Smith grants irreducible wholeness an independent status ontologically.[12] So rather than acting as a property which merely

[8] See Wolfgang Smith, "From Schrödinger's Cat to Thomistic Ontology," chap. 1, *Ancient Wisdom and Modern Misconceptions.*

[9] TQE, 118.

[10] PSQ, 15.

[11] Ibid., 19.

[12] Ibid.

accompanies something else (e.g., a substantial form) IW has real authority and in truth is the essence of all cosmic "being" besides God. After all, if there were no inherent drive to transcend a summation, how could the mere act of summation arise at all? For without such a pre-existing principle, there would be no impetus for summation to emerge—no foundation upon which it could stand. Put simply, nothing can "sum the parts" unless something already exists that surpasses both the parts and their sum. Summation, therefore, inherently points beyond itself—to something greater. To a unity that cannot be broken down: to a wholeness that is, in essence, irreducible.

With the notion of irreducible wholeness (IW) defined, Smith, in his final book, clarified that IW is the source of all VC, meaning that the two are inseparably concatenated.[13] Smith further explains that irreducible wholeness must reside within what he terms the "aeviternal realm," a domain marked by its unchangeability and situation above time.[14] In contrast, the intermediary level, which lies above the corporeal, is subject to time alone, while the corporeal realm stands subject to both space and time.[15]

Building on this framework, Smith also contends that, due to their inherent substantiality, corporeal objects are themselves instances of IW.[16] This insight convinced Smith to slightly rethink in his final book[17] how subcorporeal objects are understood. Rather than defining SX solely as the physicist's random conception of a corporeal entity X, Smith's later reformulation insists that subcorporeal objects qualify as IWs also, owing to their participation in the irreducible wholeness of X.[18] In light of this participation, Smith redefines SX as the idealized quantitative aspect of X. A quantitative aspect which by its

[13] Ibid., 25.
[14] Ibid., 21.
[15] See Wolfgang Smith, *The Tripartite Wholeness* (PSIF, 2019).
[16] PSQ, 26.
[17] Ibid.
[18] Ibid., 27.

very nature does not boil down to one quantitative description of X over another but in a sense encompasses all insofar as it is receptive to all such descriptions.[19]

At this juncture, it is also worth noting that the irreducible wholeness of an SX does not preclude it from possessing a certain degree of potency in relation to its correspondent X.[20] In Aristotelian terms, "potency" or "potentiae" refers to the potentiality of something to exist in one state or another—such as an unflipped coin landing on heads or on tails. When one of these states is realized, we say that its potency has been actualized. Within his framework of TQE, Smith argues that the potency of SX—or even of transcorporeal objects—resides in their relation to how X either presents or displays itself at the corporeal level.[21] For example, an object X might adopt a specific visual "presentation" that corresponds to a particular physical configuration at the level of its SX. Similarly, an object X may exhibit a distinct "display" at the level of its X, presenting as a measurable result on a corporeal instrument e.g., a visible pointer reading or the audible click of a Geiger counter.

The capacity for SX to accommodate various presentations and displays at the corporeal level X implies that the physics associated with SX is, to some extent, malleable. That is, multiple physical formulations can be latched onto an SX. In this sense, objects of type SX are partially intentional, but not entirely so. For there seems to exist a quantitative dimension to corporeal objects X, but the specific measurable characteristics of this dimension are, in a sense, supplied from above—by the physicist through his chosen mathematical formalism.[22]

Nevertheless, the malleability of SX does not equate to complete subjectivity. As Wolfgang Smith emphasized in his final

[19] Ibid.
[20] TQE, 74.
[21] Ibid., 35–36.
[22] Ibid, 34.

work, classical physics has its origin from the composition of an SX, whereas quantum mechanics has its origin from an interplay between IW and the transcorporeal domain.[23] Precisely how quantum mechanical entities are fashioned from the transcorporeal realm I shall explicate in a moment. Smith's rationale for this distinction lies in his metaphysical hierarchy, wherein: IW is the source of all cosmic being. Thus, a corporeal object X, which participates intrinsically in a stable IW, is naturally aligned with a physics more grounded in common sense—namely, classical physics. By contrast, the physics governing a domain with diminished ontological status—a realm further removed from IW and subject to its arrival—takes on the "spooky" character of quantum mechanics.[24] Crucially, it is my interpretation of Smith[25] that the transcorporeal origin of quantum theory and the subcorporeal origin of classical mechanics does not entail that quantum mechanics cannot be used to describe an SX. On the contrary, as we will later explore, Smith explicitly notes that quantum mechanical entities can be incorporated into an SX, albeit as "foreign entities."[26] Moreover, as Drossel and Ellis argue, the heat bath of an SX behaves classical at the global scale but quantum mechanical at the local scale[27]—a point that both Smith and I have taken to suggest that the heat bath constitutes an IW.[28] On these grounds, quantum entities can indeed be integrated into the physics of an SX, though they must retain their "origin from abroad" (i.e., the transcorporeal domain).[29] The same, however, does not hold for classical physics: although classical mechanics originates within an SX, it cannot be similarly transferred into

[23] PSQ, 30–31.

[24] Ibid., 31.

[25] Primarily based on personal conversations with the man.

[26] PSQ, 38.

[27] See "Contextual Wavefunction Collapse: An Integrated Theory of Quantum Measurement," *New Journal of Physics* 20, Nov. 30, 2025.

[28] See "Subcorporeal Physics and Vertical Causation," Part II chap. 2, PSQ.

[29] PSQ, 38.

the transcorporeal domain, precisely because the physics of the transcorporeal depends fundamentally on the interaction between the transcorporeal and irreducible wholeness.[30]

In all, we can say that the nature of objects SX can be understood as comprising three main elements. First (I), there is a fixed, irreducible wholeness that serves as an unchanging foundation, allowing X to be quantized within the framework of classical physics. This capacity is not restricted to any specific form of classical physics, thereby designating SX as an IW. Second (II), SX possesses the potential to be interpreted through various classical formalisms—be it Newtonian, Lagrangian, Hamiltonian mechanics, or others. Finally (III), SX has the ability to assimilate "foreign entities" into its quantitative corpus. More specifically, (III) reflects the capacity of SX to incorporate quantum theory—a point that will soon be clarified.

Considering all this, Wolfgang Smith uses his ontology primarily to tackle several foundational dilemmas in the philosophy of physics.

Chief among these dilemmas that Smith tackles is the famed measurement problem—the quandary of how a quantum particle can instantaneously transition from a superposition of possible states to a definite physical entity upon measurement by a classical apparatus.[31] Smith's resolution to this problem is to propose that quantum systems exist as transcorporeal Aristotelian *potentiae*. These *potentiae* are actualized in an act of instantaneous measurement by a corporeal instrument bestowing its accidental forms onto said systems.[32] The driving force behind this actualization, Smith argues, is vertical causation, a conclusion supported by the instantaneous or supratemporal nature of quantum state collapse.[33] Intriguingly, this instantaneity which is the signature of VC is not just limited to

30 Ibid.

31 Ibid., 13–17.

32 Ibid.

33 Ibid., 16.

wave function collapse but also appears across a wide spectrum of phenomena. Whether it be in the sudden "pancaking" of electric fields that puzzled Einstein or in the determination of complex specified information, instantaneous effects are pervasive in the natural sciences.[34] This fact alone underscores the reality of vertical causation which in turn smashes the materialist Weltanschauung into smithereens.

Returning to Smith's resolution of the quantum measurement problem, it is important to emphasize that the actualization of quantum potentiality does not entail the literal summoning of a microscopic particle from a nebulous cloud of probabilities. Instead, this actualization becomes manifest in the corporeal world through the perceptible outcomes registered by corporeal measuring instruments.[35] In this regard, Smith's interpretation of quantum mechanics—and of physics at large—almost bears a distinctly Carnapian overtone. That is physical phenomena correspond to phenomenal observables and receive their identity or meaning from this correspondence—something that was vigorously endorsed by the analytic philosopher Rudolf Carnap who attempted to reduce reality to sets of what he termed "observation sentences."[36]

Interestingly, this brings us to a second dilemma that Smith skillfully addresses using his ontology. Namely, the question of the ontological status of quantum particles.

In relation to their origin, quantum systems, Smith asserts, are transcorporeal entities. However, Smith argues that their generation arises primarily from the injection of IW into the transcorporeal domain—a process I have suggested is supported by the mathematics of quantum theory in its Hermitian operators.[37] As Heisenberg—himself the son of a classicist—noted,

[34] Ibid., 57.

[35] TQE, 75.

[36] See, Carnap, Rudolf; Creath, Richard & Nollan, Richard (1987). "On Protocol Sentences," *Noûs* 21 (4):457–70.

[37] See "On the Role of Irreducible Wholeness in Transcorporeal Physics," *The Disinterested Spectator* (2023).

quantum mechanics is unique among the physical theories in that the observer is not just a passive bystander but plays a constitutive role in fashioning the very phenomena under his watch. This formative involvement manifests in two key aspects of quantum theory: first, in the determination of his observational field via the selection of a specific measuring instrument and observable data.[38] According to Heisenberg this determination amounts to "posing a question." And secondly, in the collapse of the wave function that amounts to "receiving an answer."[39] Since irreducible wholeness is the source of all being Smith argues that it is through IW passing into and out of the transcorporeal domain that quantum systems are brought into existence.[40] Before this point, the transcorporeal domain contains no quantum systems, so to speak. Rather, it remains "dormant," awaiting the influx of IW necessary to bring these entities into reality.[41]

Incidentally, in certain passages of his last book Smith seems to imply that the transcorporeal realm has "no being" before the arrival of IW.[42] This description I would say is slightly exaggerated. For to have literally "no being" entails that the transcorporeal realm does not exist before the appearance of irreducible wholeness.[43] Hence, to me, it seems more correct to say that the transcorporeal realm possesses the least degree of being of any realm within Smith's cosmos. Furthermore, I suggest this is because its reality derives not from the presence of IW, but precisely from its absence. Paradoxically, it is through lacking IW that the transcorporeal realm receives its very being—granted by IW itself, albeit in a distant and

[38] See, Heisenberg, Werner, *Physics and Philosophy* (1958), chap. 3.

[39] Ibid.

[40] This corresponds to Heisenberg's process of questioning and answering.

[41] PSQ, 30.

[42] Ibid., 31.

[43] You may recall that in Scholastic parlance being is defined as essence married to existence.

indirect way, much like how a shadow receives its reality from the light it cannot hold.

Although quantum entities are indeed transcorporeal in origin, Smith also acknowledges that they play a vital role in the physics of subcorporeal objects SX. As Smith says: "the transcorporeal realm consists of *potentiae* ready to be actualized upon contact with a subcorporeal entity SX by means of VC emanating from the IW of X."[44] Smith then follows this claim up by stating that there are no quantum particles as stand-alone objects.[45] Quantum particles instead exist either as shadowy objects generated through the process of quantum measurement or as parts of an X.[46] However, while quantum systems are parts of an X, Smith makes it explicit that these systems are not parts in any reducible set-theoretic sense either. As parts, they instead contribute to the irreducible wholeness of X and in return signify X's wholeness in themselves. Quantum particles are, to put it bluntly, IWs once they have been incorporated into an X.[47]

Now if, as Smith maintains, this account is accurate, then it opens up the door to a deeper understanding of our reality—one that, I think, finds ontological support in Thomistic metaphysics.

A little-known gem of Thomistic ontology is Aquinas's theory of virtual presence. According to Aquinas, an object can be present as a part of something—not through its substantial form, but through its powers. By "powers," Aquinas is referring to the dispositional properties or capacities that an object retains.[48] For instance, in water, oxygen and hydrogen are not present in their original substantial forms; rather, they

[44] PSQ, 39.
[45] Ibid., 29.
[46] Ibid., 29–30.
[47] Ibid., 36.
[48] See Decaen, Christopher (2000), "Elemental virtual presence in St. Thomas," *The Thomist* 64 (2):271–300.

exist virtually through their inherent powers or tendencies. Their substantial forms have effectively been replaced by a new, unified substantial form: that of water itself.[49] At surface level there exist lots of similarities between Aquinas's doctrine of virtual presence and Smith's account of how quantum particles participate in the IW of an X. This suggests that a fruitful approach might be to reinterpret Smith's understanding of quantum particles through the lens of Aquinas's doctrine of virtual presence. On this reading, when a quantum system comes into contact with the IW of an X and relinquishes its individuality to become a part of X's IW, it does so by retaining the powers proper to it as a prospective individual. In other words, it sacrifices its prospective substantial form—and thus individuality—so that it may exist virtually, that is, solely in terms of its powers.

Interestingly, the introduction of Aquinas's powerist ontology offers valuable clarity on several other aspects of Wolfgang Smith's philosophy of physics. Most notably, it sheds new light on his solution to the measurement problem, which now appears far more impressive than ever before. More precisely, Aquinas's ontology finally provides a meaningful answer to how a quantum system can transition from the transcorporeal realm qua potentiality to an actual empirical reality without ever having truly existed in a concrete sense. Smith states that both before and after measurement "there are no quantum particles."[50] In light of Aquinas, this assertion now becomes far more intelligible. Throughout this process, the quantum system only ever existed virtually—that is, in terms of powers—even when it had been actualized by a measurement device.

Although Wolfgang Smith's ontology of physics is for the most part, fully developed, there do nonetheless remain a few outstanding developments. One pressing issue for me at

[49] Ibid.

[50] PSQ, 38.

least is this: if irreducible wholeness is indeed the source of all cosmic being, how then are actualizing forms imparted through Vertical Causation? For instance, in the actualization of quantum systems. The very act of bestowing forms seems to imply an act of decision-making—something that, in turn, suggests the presence of an intelligence behind that act. But what is the nature of this intelligence? Could it be the divine—namely, God? Or is it possible that such a cause could be non-divine but also intelligent? And if so, would that cause be compatible with orthodox Catholic doctrine?

While God is certainly an obvious candidate for the dispensation of forms, the idea that divine agency operates directly in their routine release, at least to me, seems somewhat far-fetched. Instead, a more compelling alternative is to hypothesize that actualizing forms are dispensed by what Plato referred to as the "world-soul."[51] For those unfamiliar with it, the world-soul—introduced by Plato in his *Timaeus*—is the animating intelligence that undergirds and unifies the entire cosmos.[52] Incidentally, the world-soul has also been entertained as a concept by holy figures in the Catholic Church, such as St Augustine.[53] In Plato's vision, the universe is not inert or mechanical, but is a living, organic whole imbued with order and purpose.[54] It is this living whole, this world-soul, that I conjecture dispenses accidental forms in the actualization of quantum phenomena. In doing so, the world-soul functions as a kind of non-divine superintelligence operating on a cosmic scale.

Moreover, the introduction of a world-soul into Smith's metaphysics provides invaluable insights into the transition

[51] The idea that hylomorphic accounts of quantum theory might be enriched through Plato's notion of the world-soul is something that has been suggested by William Simpson and Robert Koons but has yet to be applied to Wolfgang Smith's ontology specifically.

[52] *Timaeus* (34b).

[53] See Teske, Roland J. "The World-Soul and Time in St. Augustine," *Augustinian Studies* (1983), 14:75–92.

[54] *Timaeus* (34b).

from irreducible wholeness to both substantial and accidental forms. To constitute a form, after all, is to constitute an organizing intelligible principle for matter. If a cosmic world-soul exists, it stands to reason that this intelligence could mold IW into diverse principles of organization—transforming it into the wide variety of substantial and accidental forms that populate our cosmos.

May this suffice as some personal reflections on Wolfgang Smith's philosophy of physics. A man whose work has resolved not only some of the greatest ontological conundrums faced by modern science, but has also shown why science cannot have the final say about the nature of reality. The "reign of quantity" is coming to an end and in large part thanks to Wolfgang Smith!

John Taylor

Chapter V

The Metaphysics of the Measurement Problem[1]

If observation collapses the wave-function by bringing Form into contact with the *materia* of quantum potentiality, what exactly IS observation—by which I essentially mean, who is the observer? Only human consciousness and intent have the power to conceive of experiments in technical and mathematical terms, to design and construct the equipment necessary to conduct them, and to develop mathematical expressions of those experiments' results. Neither a star nor a rock nor a tree nor a cat can do any of this. Therefore, it must appear to the atheistic scientist that the human being is a kind of magician capable of transforming a mass of non-existent *potentia* into real existing substantial form by sovereign self-will alone, operating through the magical power of human consciousness. This is one of the factors behind the growth of our present magical technocracy, in which—as in the case of Jack Parsons—the technical and the magical worldviews have fundamentally coalesced. (Jack Parsons, for those who have not heard of him, was an American pioneer in rocketry who co-founded both the Aerojet Corporation and the Jet Propulsion Laboratory; he was prominent enough in his field to have a crater named after him on the dark side of the Moon. He was also a student—along with L. Ron Hubbard, founder of the Church of Scientology—of the black magician Aleister Crowley, who identified himself with the Beast of the Apocalypse. Both

[1] Excerpt from a letter to Wolfgang Smith.

Parsons and Hubbard were associated with the Los Angeles lodge of Crowley's Thelema religion. Parsons would perform pagan invocations at his rocket launchings, claimed that his life's mission was to destroy Christianity, and—as recorded in a video that may still be posted on YouTube—once presented and filmed a public demonstration of a sort of electronic ouija board he invented which, as he claimed in the video, was "how we invoke demons," his own familiar spirit being a female entity named Babalon. His life came to an end when he blew himself up in his laboratory.)[2]

However, seeing that human beings neither create themselves nor the world around them by their own action, the Human Form alone must hold the privilege of consciously participating in the same creative Act which produced it. In other words, the origin of our status as observers capable of collapsing the wave-function by introducing Form into a formless mass of *potentia* is the fact that the human being alone possesses what the Qur'an calls the *Amana*, the Trust, the role of the vicegerent of God in terrestrial existence. The same truth is manifested in Christianity by the fact that Jesus Christ, as both Son of God and Son of Man, is the epitome of the human form, "without [whom] was not any thing made that was made," and by the story from Genesis that Adam possessed the knowledge that gave him the power to name the animals in Eden—that is, to confer Form upon an originally shapeless mass of biological *potentia*. He could do this by virtue of the fact that Man is the Final Cause of the creation of the universe—this being the essence of the Anthropic Principle, which appears in Genesis in the fact that humanity is the last creation of God during His six days of world-making—as expressed in Aristotelian terms. So we can

[2] Note: I mentioned Parsons in my letter because Wolfgang, as a "rocket scientist," developed the equations that allow spacecraft to successfully re-enter Earth's atmosphere, thus making Jack Parsons a kind of "evil twin" to Wolfgang Smith.

say that the human being is capable of acting as the observer who orders and forms the corporeal world through the act of observation only because he is an object observed and thus created by the First Observer, who is God. The real creating and form-giving Observer is God alone, who surveys and so forms His corporeal and material creation precisely by observing it through the eye of Man. Plotinus taught that God creates the universe by contemplating it; therefore we must declare that God creates the human form, the epitome of universal manifestation, by exactly the same Form-giving act that the human observer uses to collapse the wave-function—because if we don't place God hierarchically above humanity as the First Observer and Creator, then Man becomes God, and the only thing that Man can be when usurping the function of God is a magician—a magician who will inevitably call upon unholy Principalities and Powers to back him up, just as Jack Parsons did. In other words, if we don't extend the axis of vertical causality all the way up to the Logos itself, we will commit the Promethean transgression.

This vertical causality is graphically portrayed by William Blake in his famous images.[3] The first shows the physicist (Newton) introducing order and measurement into the world of *potentia*; the second shows God creating the terrestrial universe by exactly the same method, and in so doing weaving the veil of *Maya*, a word derived from the Sanskrit root *ma* (cognate with the Greek *meter*) meaning "to measure." The physicist introducing measurement into the chaos of material *potentia* and the figure of the God creating the corporeal and material universes are shown as *bent over* to indicate that both the sensual corporeal world and the sub-sensual material world, though they are certainly real, are created through a *contraction* of consciousness and a limitation placed upon the

[3] You really must see these paintings: *Newton* (1795), color print in pen and ink and watercolor, and *The Ancient of Days* (1794), ink on paper, both at https://www.posterlounge.fr/artistes/william-blake/.

fullness of Being. Likewise, *Maya* is not pure illusion—as the Hindus traditionally describe it, it is "like a rope mistaken for a snake"—but a radical *editing* of the fullness of Being by the process of measurement, which only becomes an actual illusion when we mistake it for all that is.

The creation of the world conceived of as the union of *forma* and *materia* to produce Substantial Form was not invented by Aristotle; it is clearly laid out in the first chapter of Genesis: "In the beginning, God created the heavens and the earth" (*forma* and *materia*). "And the earth was without form, and void, and darkness was upon the face of the deep" (terrestrial existence was originally mere *potentia*). "Then the Spirit of God moved upon the face of the waters" (the Spirit being the breath of the creative and form-giving Word). In the first five days of creation the union of *forma* and *materia* was incomplete; it was only finished on the Sixth Day with the creation of the Human Form, who bears the *imago dei*, and is therefore Substantial Form complete—the one who, alone of all creatures, possesses the power to participate in God as Pure Act, Who is the perfect actualization in Eternity of the universal *potentia*. As for the Seventh Day, when God rested from His labors, this represents His activation of the contemplative function in Man; the power that allows us to see creation whole, and therefore to contemplate its Creator; the faculty that frees us from the limitations of dimensional, created existence and sets us on the path of return to our Divine Source.

Charles Upton

Chapter VI

The Measure of Things

Quantum Physics and the Rediscovery of Meaning

Introduction

Wolfgang Smith was a philosopher in the true sense of the word. He loved wisdom and sought understanding above all. He combined rare intellectual range with a lifelong search for truth. Born in 1930, he completed three degrees from Cornell—mathematics, physics, and philosophy—by the age of eighteen. He went on to earn a master's in physics from Purdue and a PhD in mathematics from Columbia. After completing his studies, he contributed to the spaceflight reentry problem and held academic posts at MIT and UCLA before eventually settling at Oregon State University, where he could balance teaching with retreating to his mountain home to study Thomas Taylor's translations of Plato.

Other pivotal moments in Smith's life included escaping World War II through Russia as a child, meeting Albert Einstein as a boy, and traveling through the Himalayas as a young man in search of the wisdom of the sadhus he first encountered at Cornell in the poetry of Rabindranath Tagore. Eventually, he met and married his beloved wife Thea, also a mathematician and seeker. It was through the "silent and spiritually uplifting influence of Thea" that Smith revisited his Christian roots and deepened the ontological vision that would define his life's work.

By the late 1940s, when Smith entered Cornell, Schrödinger's wave mechanics had been established for two decades, and physics was widely regarded as the philosopher's key to reality. Asked why he chose physics, Smith answered simply: "*because I believe that physics is the key to understanding the universe.*"[1] Yet, as he later reflected, the journey turned out differently:

> Well, I no longer think so, and have in fact become persuaded that the matter stands just the other way round: that in order to understand physics, one needs first to attain a certain insight concerning the universe—an ontological insight, to be precise.[2]

That reversal—beginning with physics as gateway but finding in it a deeper pointer toward ontology—captures the trajectory of his thought. Physics never supplied the final answer, but it left traces, signs which, when read together, suggested a startling conclusion: *the cosmos is ordered not only by number but by meaning.*

This early search for answers through physics would lead him to insights he would later distill when reflecting back on his first book:

> What I object to fundamentally in the scientistic worldview is that it conceives the objective universe to be unperceived and unperceivable.[3]

> The decisive fact is that the cosmos does not stand alone: it cannot in truth do so even for an instant.[4]

This essay follows Smith's path: to trace how his critique of scientism, his categories of corporeal, subcorporeal, and transcorporeal, and his Christian vision of the cosmos as icon

[1] Preface to the First Edition of *Physics: A Science in Quest of an Ontology*, xi.

[2] Ibid.

[3] Preface to the Third Edition of *Cosmos & Transcendence*, ix.

[4] Ibid., x.

together illuminate why physics itself discloses, not a world of brute quantities, but a world suffused with meaning.

We will begin with Smith's basic ideas before turning to the act of classical measurement (the desk), quantum measurement (the electron), and the origins of quantization (Planck). From there we will weigh modern interpretations of quantum mechanics, consider why the observer tends to vanish, and then turn to the longer cultural tradition—above all, the Church—that cultivated the grammar of sign and signified without which science itself could not have taken root.

Some technical details of physics are unavoidable, but we will keep them as simple as possible. Readers not interested in equations can skim those passages and still arrive at the heart of Smith's insight: that the cosmos is not closed upon itself but ordered by number and meaning, sign and signified, the logos of reason and the Logos of the divine Word.

Science and Scientism

It is easy to mistake Smith for a critic of science, but science itself was never his target—he loved it, celebrated its precision, and taught it at the highest level. What concerned him was a metaphysical assumption that often trails behind science unnoticed: the belief that quantitative measurement is the sole source of objective truth and that only what can be measured can be regarded as real.

When we talk about what is real, we are talking about ontology: the branch of philosophy that asks, *what kinds of things exist, and how do they relate to one another?* Once central to philosophy, ontology has largely fallen out of fashion, displaced by narrower technical questions. For Smith, however, ontology was not antiquated. He saw himself as an ontologist in the classical sense—one who studies the nature and the structure of reality itself.

By the twentieth century, the neglect of ontology in physics, in Smith's view, had become a crisis. He often quoted physicist

Nick Herbert, who remarked that modern physics had "lost its grip on reality." By this he did not mean physics had lost rigor, but that it no longer seemed to tell us what the world is—only what numbers to expect on our instruments.

For Smith, this was not a failure of physics but a reminder of what physics actually is: the science of measurement. Its great power lies in quantifying relations with astonishing precision. The problem arises when we assume that only what can be measured is objective reality, and that what cannot be measured is somehow unreal. Scientism, in other words, collapses sign and signified. Numbers and models, instead of being bridges to reality, are mistaken for the reality itself.

To help us understand this forgetfulness, Smith distinguished between two worlds. The corporeal world is the world of lived perception: the coffee mug on the desk, its color, its warmth, its presence as a whole. The physical world is the world as seen by the physicist through measurement and equations—a world of particles, forces, and fields, stripped of qualities to emphasize magnitudes.

This may remind some readers of Descartes' famous division between *res cogitans* and *res extensa,* mind and matter. But Smith's point was different. He was not reopening a dualism but correcting it. The physical world is not the whole of reality but one layer within a multi-leveled order of being, alongside the corporeal world we actually inhabit.

Many readers will find Smith's ideas hard to accept, and understandably so. They cut against more than four centuries of Western habit. Since Descartes, we have been trained to think of the "real world" as the world of equations and magnitudes, while the world of lived perception is dismissed as "mere appearances." I can say from experience that this was not easy for me either. I hold a STEM PhD and spent two years in weekly conversations with Wolfgang Smith himself. Yet even then, I struggled to grasp what he really meant by the corporeal versus the physical.

Only slowly did I come to see how Smith encourages us to neither lose the corporeal in the physical nor the physical in the corporeal. Both are real, but they are distinct. His aim is to recover a more traditional and layered understanding of reality while fully preserving the integrity of science. Far from dismissing science, he sought to protect it. For scientism, he warned, ultimately threatens science itself. By reducing the real to only what can be measured, scientism undermines the very basis of intelligibility: the conviction that the world is meaningful and therefore worth investigating in the first place.

Measuring the Desk

Rather than jumping straight into electrons and quanta, we begin with something deceptively simple: measuring the width of a desk. Stretch a tape measure across it: *29¼ inches.* A gesture anyone can replicate, yet already layered with meaning.

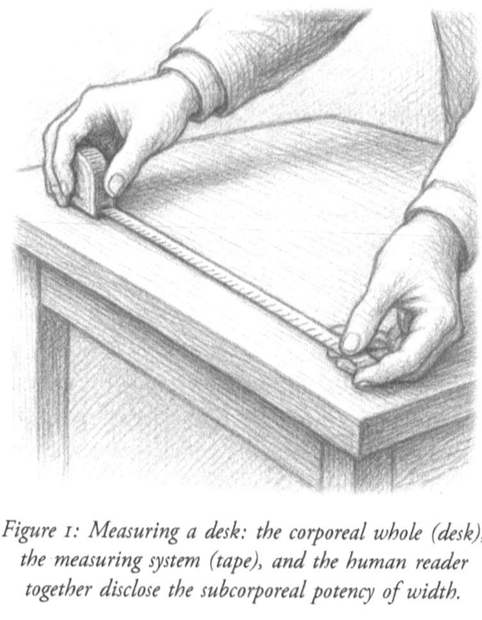

Figure 1: Measuring a desk: the corporeal whole (desk), the measuring system (tape), and the human reader together disclose the subcorporeal potency of width.

Every measurement involves three elements: the object measured (the desk), the device (the tape), and a person who performs the act and perceives the result. Desk, tape, and numerals all belong to Smith's first category, the corporeal—entities directly perceivable without theoretical mediation.

But what about the "width" itself? On one level, you can directly perceive it: the expanse of the desk is visible, traceable by hand, even embraceable by outstretched arms. That is the corporeal reality. Yet the number *29¼ inches* is not simply lying in the wood. It arises from another layer of the physical world, which Smith called the *subcorporeal*: an invisible dimension of quantifiability tied to corporeal objects. The number is not another object beside the desk; it comes into view only in the act of measurement.

The key feature of these classical potencies is their *determinacy*. The desk really does have one definite width, whether or not anyone measures it. This hidden but determinate dimension—the subcorporeal—becomes actualized when person, desk, and tape are brought into relation. Measurement does not create width *ex nihilo*; the potency is already fixed in the corporeal desk. What measurement does create is the visible numeral—the sign—which discloses that potency in readable form.

Psychologist James J. Gibson's research on visual perception is suggestive here, and worth noting because of its importance to Smith. Where many theories treat perception as construction—an internal model assembled by the brain from raw sensory data—Gibson argued instead that perception is disclosure. The world presents affordances, real but invisible invitations like graspability or climbability, which become actual in action. For Smith, this aligned closely with the subcorporeal dimension. Just as a desk's quantifiable width exists as a determinate potency awaiting measurement, so too an affordance exists as a determinate possibility awaiting enactment. Gibson's work thus reinforced Smith's conviction that perception is disclosure, not construction.

The number itself—the visible "29¼ inches"—is a *sign*. What it signifies is the potency of quantifiable width, invisibly anchored in the desk but not identical with it. The sign is never the signified. Scientism collapses the two: it assumes that the number is the width.

This may seem trivial at the scale of a desk, but consider Newton's laws of motion. The signs were corporeal: the observed positions of Mars, painstakingly recorded by Tycho Brahe. The signified was the hidden potency of orbital order, which Kepler and Newton described. Newton's equations let astronomers predict Mars's position decades ahead. But the laws did not cause Mars to move; they were grammar mediating between observed marks and invisible potencies.

Here the temptation of scientism becomes stark: to treat the model as the ground of reality itself. The "inverse-square law" is spoken of as though it were the force, rather than a formalism mediating between visible signs and invisible potencies.

Smith resists this flattening. He insists that signs remain signs, and that invisible potencies remain distinct. In classical science, the distinction was easy to overlook because potencies were determinate and repeatable. Promissory materialism—the hope that finer instruments would someday close the gap—kept the illusion intact. But quantum mechanics would shatter that illusion.

Measuring the Electron

If the desk showed us the classical pattern—an invisible but determinate potency anchored in a visible whole and disclosed by a visible sign—the electron shows us something new: an invisible and indeterminate potency that becomes determinate only in the act of measurement, where it is also disclosed by a visible sign.

Consider the familiar double-slit experiment. A stream of electrons is directed toward a screen with two slits. On the far side, a detector records where each electron lands. What we actually observe are dots—corporeal marks on the detection screen. The electron itself is never seen; only the marks appear.

The theory, however, describes something stranger. Each trial yields only a single dot, yet the model insists that what is "really happening" is a wave-like superposition spreading through both slits at once. The model speaks in smooth probabilities, while the apparatus delivers discrete marks. Only after hundreds or thousands of runs does the statistical pattern of the dots resemble the interference fringes predicted by the wave model. This mismatch—between continuous formalism and discrete corporeal signs—is the hallmark of quantum measurement.

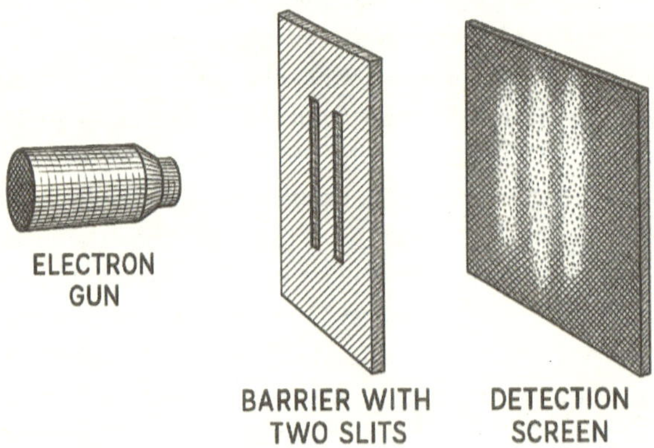

Figure 2. Quantum measurement: the electron itself remains unseen; the only corporeal sign is the dot registered on the detection screen.

The contrast with the desk is stark. The desk's width belongs to the subcorporeal: a hidden but determinate potency of a corporeal whole. The electron resists such anchoring. Its "where" or "how fast" are not merely concealed properties waiting to be revealed; they are indeterminate until the act of measurement.

Here Smith clarifies a further level of the physical, distinct from the subcorporeal: the transcorporeal. These are not invisible potencies tied to visible objects. Rather, they are abstract formal structures—wavefunctions, Hilbert spaces, probability amplitudes. They organize what is possible, but do not themselves belong to the corporeal order. They live in the space of theory, yet they constrain what corporeal marks can appear in quantum measurements.

The three levels can now be seen together:

> *Corporeal* (X): directly perceivable formed objects, including the visible symbols that terminate measurement.
>
> *Subcorporeal* (SX): imperceptible but determinate quantifiable potencies anchored in corporeal wholes.
>
> *Transcorporeal* (T): entities described only by equations—wavefunctions, Hilbert spaces, probabilities—not directly perceivable, yet indispensable for predicting the distribution of corporeal marks.

Smith's notation is instructive here. Corporeal objects are symbolized by X. Subcorporeal entities are symbolized by SX to indicate that a subcorporeal potency SX is always tied to a corporeal object X. A transcorporeal entity T, on the other hand, is not denoted TX, because it is not anchored to any corporeal whole. The desk's width (SX) exists because it is tied to the desk (X). The electron's wavefunction is not the hidden property of a visible object—it is a structure of possibility itself.

Quantum measurement does, however, still involve an X and SX—just not of the electron. In the case of electron position, the X is the detection screen, and SX is the potency of possible positions on that screen. The dot you see is the screen's potency becoming act, constrained by the electron's wavefunction. A look back at Figure 2 highlights that the only observable marks in the double-slit experiment are the marks on the detection screen. No one ever observes the electron in any way whatsoever, other than the mark on the detection screen.

The similarity with the desk shows continuity: measurement always terminates in a corporeal sign. The difference shows the novelty: the ruler's number discloses what was determinate all along, while the electron's dot discloses what becomes determinate only in the act of measurement.

With the desk, measurement reveals what was already fixed; with the electron, measurement participates in making it definite. Quantum mechanics inaugurates a new pattern: the invisible remains indeterminate until the very act of measurement, where it is transfigured into a determinate mark.

And yet, across both cases—desk and electron alike—the structure of science converges at the same place: a corporeal sign. Whether a number inscribed on a ruler or a dot glowing on a detection screen, every act of science terminates in the visible mark.

Planck's Problem: The Birth of Quanta

This reliance on corporeal marks is what pushed physics, at the dawn of the twentieth century, into uncharted territory. In the study of radiant heat, the smooth experimental spectra could not be reconciled with smooth theoretical models.

Experiments on "black bodies"—cavities heated until they glowed—produced beautifully continuous spectra of colored light. These spectra were the signs: corporeal marks of brightness at each wavelength, carefully recorded. The signified, according to classical theory, was continuous: oscillators in the cavity walls exchanging energy smoothly. But the signs refused to line up. Classical equipartition demanded an infinite rise of energy at short wavelengths—the "ultraviolet catastrophe."

Planck's "desperate act" was to sever continuity. He assumed oscillators could exchange energy only in discrete packets, quanta proportional to frequency. With one stroke, the infinities vanished, and the signs matched the model. At first he treated this as a calculational trick, but in hindsight we see what it disclosed: not flawed instruments but a new phenomenon of nature. The corporeal marks could not be reconciled with continuity. Reading them truthfully required acknowledging discreteness. And once that became clear in the quantum case, it cast retroactive light on the classical case as well: what had seemed like smooth continuity was itself already a disclosure through determinate potencies, not an absolute transparency of nature.

Importantly, this discreteness was not universal. Quantization reveals itself only under specific conditions:

1. Systems with discrete allowed states (oscillators, bound electrons).
2. Energy exchanged between those states.
3. Apparatus sensitive enough to record individual events as irreversible marks.

When these conditions are met, quantization is not just a modeling device but an ontological disclosure. Einstein confirmed it with photons, Bohr with atomic jumps, de Broglie with matter waves, Schrödinger with quantized energy levels. What began as Planck's *ad hoc* fix became recognition of a real order of being.

Here lies a temptation. Once discreteness is found in one corner, scientism urges us to flatten all of nature into quantization. But Smith shows why this is a mistake. Quantization is not a universal principle; it is a layer of being disclosed under particular conditions. Outside those conditions, nature remains continuous in its mode of disclosure.

Quantum mechanics did not create the gap between sign and signified; it made it impossible to ignore. What collapsed was not the wavefunction but the illusion—nurtured by scientism—that horizontal refinement alone could bridge sign and signified. What replaces collapse is what Smith calls *vertical causation*: the ontological passage from potency to act.

Modern Interpretations and Ontological Temptations

Once quantization was accepted, physicists faced a deeper question: how to interpret what the equations meant. Three schools of thought emerged, each flattening the distinction in its own way.

The *Copenhagen interpretation* treated the wavefunction not as reality but as a predictive tool. Collapse was a bookkeeping move, a rule for updating expectations. The temptation here is to collapse sign into signified, flattening the transcorporeal into mere calculation.

Many-worlds, by contrast, removed collapse entirely. Every possible outcome actually happens in parallel universes. The dot you see is real, but so are all the other dots you did not see. The temptation here is to preserve horizontal causation by multiplying worlds, erasing vertical causation: the step by which potency becomes act.

Hidden-variable theories, favored by Einstein and later Bohm, insisted that quantum indeterminacy must conceal deeper determinism. The dot on the screen was treated as if it only masked a determinate potency always there. The temptation here is again to collapse sign and signified. Yet experiments—Bell's theorem and Alain Aspect's tests—showed no such hidden layer could exist.

Despite their differences, these interpretations share a common flattening. Some collapse sign into signified, others deny collapse, still others posit hidden determinacy. In each, the observer fades out. Either equations suffice, or parallel worlds take the burden, or hidden variables do the work.

Smith resists all three. He insists corporeal, subcorporeal, and transcorporeal are ontologically distinct; that the act of measurement is not an embarrassment but the site where they meet; and that the human measurer is integral, not dispensable. For Smith, the real collapse is not of the wavefunction but of scientism's illusion that horizontal causation alone could explain measurement.

The Vanishing Observer

Compare measuring a desk with physicists' depiction of measuring an electron. In the first, the human role is explicit: a person stretches the tape, reads the scale, and records the number. The corporeal sign—"29¼ inches"—is inseparable from this embodied act.

In the electron case, the person vanishes. Diagrams show an electron gun, slits, and a detection screen. Dots appear as though generated automatically, as if the marks read themselves. Yet in reality, every experiment requires someone to receive and record the disclosure.

Why does science excise the observer? Four reasons converge:

1. *The pursuit of objectivity.* Science promised knowledge that would hold regardless of who observed. Marks detached from persons seemed more secure, more universal.
2. *The Cartesian bifurcation.* Descartes split reality into res extensa (extended matter) and res cogitans (thinking substance). Science claimed the former and bracketed the latter: what was measurable was real; what was conscious, incidental.
3. *The power of equations.* From Newton onward, mathematical laws became the grammar of science. Their elegance lay in erasing particularity. Schrödinger's wavefunctions, like Newton's laws before them, speak as though no one were there to read the signs.
4. *The allure of autonomy.* Instruments appear to "read nature directly." Yet the numbers they yield are never self-interpreting; without someone to receive them, they remain marks on paper or blips on a screen.

This exclusion is not merely artistic convention. It is built into the migration from observed marks to abstract formalisms. The observer receded—from notebooks, to background, to equations.

For Smith, this was more than a sociological drift; it was a metaphysical error. Every measurement requires a perceiver. Every corporeal mark is a sign that must be read. To omit the human is to omit the very act by which potency becomes act, invisibility becomes visibility.

The Collapse of Scientism

Every act of science, from desks to electrons, terminates in the same place: a corporeal sign. A number written, a dot on a screen, a trace on a plate—without these visible marks, there is no science. They are anchors of knowledge, the meeting point of potency and act.

But marks never interpret themselves. They are read through models—formal grammars mediating between seen and unseen. Some map determinate potencies (subcorporeal), others indeterminate possibilities (transcorporeal). In every case, models are bridges, not terrain.

To collapse sign and signified—to mistake model for reality—is scientism. In classical science, this collapse could pass unnoticed because potencies were determinate. In quantum physics, it is impossible: the dot is not revelation of hidden determinacy but the very event by which potency becomes act. Quantum mechanics unmasks what was always true: science lives by distinction.

Here Smith's categories clarify: corporeal, subcorporeal, transcorporeal are distinct layers, joined not by horizontal continuity but by vertical causation. Forget this, and we reify models, multiply worlds, or deny collapse. Remember it, and the puzzles fall into place.

From Collapse to Custodianship

What remains is a cultural question: how do we preserve the distinction between sign and signified? How did we arrive at it, and how do we resist its collapse?

From her earliest days, the Church insisted visible realities are not the whole. Her sacramental vision trained the imagination to see visible things not as idols closing in, but as icons opening onto what they signify. Bread and wine are not "mere" bread and wine; they disclose a deeper reality. This was ontology: the world is composed of signs that both reveal and conceal, that must be interpreted, and that participate in what they disclose.

By preserving this grammar of sign and signified, the Church safeguarded personhood. Against empires that reduced people to instruments, she proclaimed each human as *imago Dei*—a visible being signifying the invisible God. Against philosophies that dissolved reality into abstraction, she upheld creation as radiant with meaning.

This grammar became the seedbed of science. Medieval universities—nurtured by the Church—cultivated the disciplined reading of the Book of Nature alongside Scripture. Measurement, mathematics, experiment could flourish because the world was open to meaning.

When Copernicus, Galileo, and Newton mathematized nature, they did so within this cultural imagination. Kepler called himself a "priest of the Book of Nature." Galileo said the book was written in mathematics, but never mistook numbers for reality itself. Newton regarded laws as signs of divine order. The theological soil nourished the scientific tree.

Yet the temptation to collapse sign and signified never disappeared. As theology receded, science inherited the mediating role. Laws hardened into forces, models into ontology, and the cosmos itself was treated as an idol: closed, self-sufficient, governed only by horizontal causation.

The Cosmic Icon

Here Smith reenters the story: to remind us of what has been forgotten. His hierarchy sharpens the decisive fact: "the cosmos does not stand alone."

By "cosmos," Smith means the ordered whole of reality as disclosed to us—a world measurable, perceivable, and meaningful. The cosmos includes stars and particles but also color, warmth, beauty, significance.

For Smith, the cosmos is not an idol but an icon. It is not closed but open, transparent to what transcends it. This is not metaphor but ontology: visible realities signify and participate in invisible orders.

An idol-based cosmos expects every puzzle—including the measurement problem—to resolve within equations. An icon-based cosmos reframes it. The wavefunction does not collapse mathematically because the transition is vertical: from potency to act, invisible to visible. Disclosure requires a participant.

Smith pictured this as a circle: the center is the transcendent source, the circumference the corporeal world, the interior the invisible potencies, the radii the acts of disclosure.

In this geometry, the human person is indispensable, spanning all dimensions: corporeal in body, subcorporeal in perception, transcorporeal in intellect. Measurement is not puppeteering nor passive reception but reciprocity: the human intends, the world responds with a sign, and the human looks back through it toward intelligibility. The cosmos, in this light, is a cosmic icon: a reality that reveals itself and draws the mind beyond itself, a window onto the Logos.

The Quantum Catholic

Now we can see what Smith meant when he said that to understand physics one must first attain ontological insight. His own relationship with physics became an illustration of the cosmic icon: physics revealed order in number, but in its intelligibility pointed him back to Intelligence itself.

This is why it is fitting to see Wolfgang Smith as the Quantum Catholic. He refused the false rivalry between science and faith. Physics was not theology's enemy nor its replacement, but a gateway—a sign pointing back to the Logos. What scientism distorted was precisely this: the temptation to mistake sign for signified, number for reality, cosmos for idol, physics for Logos.

Against that distortion, Smith restored the conviction that the cosmos is not self-sufficient matter but a layered reality, ordered by number and meaning. His categories were not taxonomy but Christian metaphysics: vertical causation in which signs participate in what they signify.

He did not merely interpret theology through physics or physics through theology. He inhabited their union, showing that the decisive fact—that the cosmos does not stand alone—was not abstraction but lived reality: science and sacrament, logos and Logos, disclosure and transcendence joined.

Logos and Communion

Smith's invitation is clear: recover a vision in which science and religion, measurement and being, stand not in rivalry but in communion.

In the Church's sacramental vision, that union was already given and became the soil in which science grew. Signs were not substitutes but participations. Matter was radiant. Measurement itself became mediation. Physics disclosed number; the Church disclosed meaning. Together they formed a communion of intelligibility and Intelligence—logos and Logos.

Smith once wrote:

> What an unspeakably joyful revelation it is to discover that—in place of raw matter wandering through space to no purpose—the world is in truth replete with sound and color, as well as a host of other sensible qualities: that it is in fact the very world artists have painted and poets have sung... inhabited by beings we can perceive, communicate with, and love.[5]

This brings us back to the beginning. Physics was Smith's gateway, but it pointed beyond itself. The cosmos does not stand alone. Every equation, every experiment, every mark testifies not only to number but to meaning.

To say the cosmos is ordered not only by number but by meaning is not to diminish science but to restore it. Number

[5] *Three Scientistic Heresies*, 6.

discloses order; meaning discloses intelligibility. Together they point beyond themselves to the Logos in whom they subsist.

This, at last, is the true measure of things: science restored to wonder, religion restored to mediation, the small-l logos restored to the Logos.

And yet we know, with Augustine, that such recognition often comes late. "Too late have I loved thee," he confessed, awakening to the beauty that had always already sustained him. Our culture finds itself in a similar posture: belatedly realizing that in losing the grammar of sign and signified it has imperiled not only faith but science itself. The invitation of Wolfgang Smith is to recover what has been there all along—that the cosmos does not stand alone, but shines as icon, radiant with meaning, opening onto the Logos.

Epilogue: The Search for a Stable Center

If Smith's insights are to matter today, they must also address the cultural mood—a generation oscillating between longing for sincerity and suspicion of all grand narratives. This cultural oscillation is often described by the term *metamodernism.* Metamodernism inherits the postmodern critique, which exposed the fragility of grand narratives and unmasked the hidden power of institutions. Yet unlike postmodernism, it does not rest content with irony or deconstruction. It longs for coherence, for sincerity, even for the sacred—while remaining acutely aware of contingency.

In this sense, metamodernism sees science not as an unquestionable arbiter of truth but as one grand narrative among others, subject to the same fragility and suspicion. At the same time, it yearns for the stability and meaning that once animated premodern and religious worldviews. It is a restless quest: a search for meaning that recognizes both its necessity and its provisionality.

From this vantage, the Catholic Church often appears the problem: the ultimate grand narrative, the most rigid institution, the embodiment of dogmatic certainty. For the metamodernist, she looks like a relic incapable of addressing pluralism and flux.

Yet this misses something crucial. It sees the Church only as human institution, judged by the same metrics as others. It fails to recognize her ontological role: custodian of the grammar of sign and signified. What the metamodernist seeks—a center that avoids both rigidity and relativism—is precisely what the Church has carried for centuries.

Catholic dogma, from this perspective, is not brittle rules but a robust generative model. It integrates physics and metaphysics, body and soul, history and eternity. The Trinity is the ultimate relational ontology. The Incarnation is vertical disclosure of transcendence. The Eucharist is symbolic disclosure par excellence, where bread and wine participate in what they signify. Dogma here is not the enemy of freedom but its precondition: a stable center from which life can unfold without collapse into chaos.

The longing for sincerity, haunted by rivalry, finds its cure in the Paschal Mystery. As René Girard showed, our desires are mimetic, easily trapped in cycles of rivalry resolved by scapegoating violence. Christ, the innocent victim, unmasks and defeats this cycle by offering himself in non-rivalrous love. His sacrifice grounds a sincerity no longer naive, because it rests not on fragile self-assertion but on communion with the divine.

In this light, the Church's claim to truth is not oppressive but liberating—the semantic foundation that makes all meaning coherent.

Yet there is also a warning for the Church. She faces her own temptation of collapse: to mute her distinctives to appear more acceptable, or to treat her signs as self-contained. If she abandons her role as custodian of sign and signified, she abdicates the gift she alone can offer. Her mission is not to mirror cultural oscillations but to hold open the space of meaning—to preserve the icon from becoming an idol.

The invitation, then, is twofold. To the metamodernist: recognize the Church not as enemy of sincerity but as its possibility, the stable center where meaning discloses itself without collapse. To the Church: remember your vocation, and resist the temptation to dissolve into cultural categories. For only by remaining faithful to her grammar can she continue to be what the world most seeks: a true and lasting home for meaning.[6]

Richard Smith

[6] This essay was completed on the Feast of Our Lady of Ransom, September 24, 2025. May it stand as a small act of gratitude for her intercession and maternal protection.

Chapter VII

Interpretation of Quantum Mechanics: *top-down or bottom-up ontology?*

> There is no higher law than that of truth.
>
> Frithjof Schuon

Wolfgang Smith (1930–2024), a distinguished mathematician, physicist, and philosopher, offered a profound re-evaluation of quantum theory through the lens of Perennial Wisdom. In the preface to his seminal work *The Quantum Enigma*, he introduces the central theme of his inquiry:

> This book is about quantum physics, or what has been termed the "quantum reality" problem. It deals with an enigma that has tantalized physicists, philosophers, and an ever-widening public for decades. The pertinent literature is vast, and it would appear that just about every conceivable avenue of approach to the problem—no matter how seemingly farfetched—has been advocated somewhere and explored. Gone are the days when the authority of physics could be invoked in support of a single established world-view! What has happened is that the pre-quantum scientistic world-view (now termed "classical") has come to be disavowed "at the top": by physicists capable of grasping the implications of quantum theory. And this in turn has called forth an abundance of conjectured alternatives, competing with one another, as it were, to fill the ontological void—a situation that has prompted one recent author to speak of a "reality marketplace." Quantum mechanics, if you

> will, is a scientific theory in search of a *Weltanschauung*. The search has been on since 1927.[1]

Smith continues:

> The new physics came to birth during the years 1925 and 1926. By the time physicists gathered at Como in 1927 for the International Physics Conference, the foundations of nonrelativistic quantum theory had been laid. It was later during the same year, when physicists met again in Brussels for the Fifth Solvay Conference, that the quantum debate erupted in full force, so to speak, in the form of the celebrated Bohr-Einstein exchange.

And he warns:

> Meanwhile the spectacle of a dozen top-ranking scientists promoting twelve different world-views is hardly reassuring; and there is the temptation to conclude that truth is unattainable, or, worse still, that it is relative, a matter simply of personal opinion.

In the following section, I aim to provide a concise overview of these competing interpretations of quantum mechanics, a theory that has persisted for nearly a century. Then, I will outline how Wolfgang Smith's interpretation not only diverges from these views but offers a compelling metaphysical alternative grounded in vertical ontology—a downward rather than upward vision of reality.

Different Interpretations of Quantum Mechanics

Interpreting quantum mechanics involves attempting to explain how the mathematical formalism of the theory corresponds to physical reality. Quantum mechanics has undergone extraordinarily rigorous experimental testing across a vast range of

[1] *The Quantum Enigma: Finding the Hidden Key* (Hillsdale, NY: Sophia Perennis, 2005).

phenomena. Remarkably, no empirical observation to date has contradicted its predictions. Wolfgang Smith highlights this extraordinary harmony between theory and experiment:

> Mention should be made, in this connection, of its so-called static attributes, such as mass, charge, and spin. Unlike the dynamic attributes—which, as we have seen, are not attributes at all—these quantities do belong to the electron as such. And they are measurable with stupendous accuracy. Recent measurements of the magnetic moment, for instance, have led to the value 1.001 159 652 188 (in appropriate units), with a possible error of 4 in the last digit. As Richard Feynman has pointed out: "If you were to measure the distance from Los Angeles to New York to this accuracy, it would be exact to the thickness of a human hair." Moreover, this magnetic moment can also be calculated by way of quantum electrodynamics; the answer appears then as the sum of a convergent infinite series, in which the successive terms decrease rapidly but become progressively more laborious to evaluate. And whereas calculations completed up to this time have not yet been able to match the accuracy of the latest experiments, they have indeed confirmed the digits 1.001 159 652. We know of no domain of physics in which agreement between theory and experiment has been more spectacular.[2]

Despite this astonishing agreement between theory and experiment, numerous schools of thought continue to contend over how to interpret quantum mechanics. These interpretations differ on key philosophical issues, such as whether the theory is fundamentally deterministic or indeterministic, what constitutes physical reality, and how the process of measurement should be understood. Such issues are also entangled with broader questions, like why time seems to flow in a

[2] Ibid.

specific direction and how the "classical" world emerges from quantum foundations.

Even after nearly a century since its inception, no consensus has emerged among physicists and philosophers regarding which interpretation offers the most accurate account of reality. Indeed, the meaning of some foundational concepts in quantum theory has evolved. For instance, Erwin Schrödinger (1887–1961) initially conceived of the wave function as representing a spatially extended charge distribution of the electron, while Max Born later reinterpreted it probabilistically—as the square of a probability density function.

Quantum mechanics can be mathematically formulated in several equivalent ways, the most prominent being Schrödinger's wave mechanics, Heisenberg's matrix mechanics, and Feynman's path integral formulation. Most interpretations accept the same formalism but diverge on how to relate it to observed phenomena. For example, the pilot-wave theory, developed by Louis de Broglie (1892–1987) and later refined by David Bohm (1917–1992), introduces an alternative mathematical model aimed at resolving conceptual paradoxes. Yet because these alternative approaches yield nearly identical empirical predictions, they are typically viewed not as competing theories but as interpretations of the same underlying framework.

A more nuanced perspective suggests that each interpretation subtly alters the theory by introducing new theoretical elements. In this sense, interpretations may be more than mere commentaries—they may constitute distinct conceptual frameworks built on shared mathematical scaffolding.

Broadly speaking, interpretations fall into two categories: epistemic and ontological. An epistemic interpretation sees the probabilistic nature of quantum mechanics as stemming from human limitations in knowledge or measurement, while an ontological interpretation regards it as an intrinsic property of reality itself. These viewpoints are grounded in broader philosophical stances. Scientific realism holds that scientific theories aim to describe the world as it truly is—both epistemically and ontologically—whereas anti-realism is more circumspect.

In the early twentieth century, the dominant anti-realist view was logical positivism, which aimed to eliminate references to unobservable entities from scientific discourse. From the 1950s onward, anti-realism evolved into instrumentalism, which permits discussion of unobservables but maintains that theories are merely tools for prediction rather than windows into reality. The instrumentalist ethos is encapsulated in the now-famous quip by physicist David Mermin: "Shut up and keep calculating."

Copenhagen Interpretation

The Copenhagen interpretation, still accepted by many physicists as the standard interpretation of quantum mechanics, was developed by Niels Bohr (1885–1962) and Werner Heisenberg (1901–1976) during their collaboration in Copenhagen around 1927. It builds on Max Born's (1882–1970) probabilistic interpretation of the wave function. According to a poll conducted at the 2011 conference on *Quantum Physics and the Nature of Reality*, the Copenhagen interpretation received the highest support, with 42% of participants favoring it. However, Nobel laureate Steven Weinberg (1933–2021) noted in a 2017 article that "the Copenhagen interpretation is now largely unacceptable."

The main principles of the Copenhagen interpretation are as follows: Every quantum mechanical system is represented by a wave function, which contains all the information that can be known about the system. There are no "hidden variables" that must be added to complete the theory. In non-relativistic quantum mechanics, the evolution of the wave function over time is governed by the deterministic Schrödinger equation—so long as the system remains isolated. As such, the wave function changes continuously and predictably, unless disturbed by interaction with another system.

Certain pairs of observable quantities cannot be simultaneously determined with arbitrary precision. This idea is central to Bohr's principle of complementarity and is quantitatively expressed by Heisenberg's uncertainty principle. For instance, if the exact position of a particle is known at a given moment, its momentum cannot be precisely determined at that same moment. Quantum processes at atomic and subatomic scales are inherently inaccessible to direct observation because any act of observation significantly alters the system being observed. Generally, as the number of particles in a system increases, the system's behavior begins to resemble that predicted by classical physics. Bohr argued that classical concepts are conceptually independent of quantum theory and retain a foundational role.

The Copenhagen interpretation dismisses questions like "Where was the particle before the measurement?" as meaningless. According to this view, measurement selects one of the possible eigenstates described by the wave function, based on the probabilities encoded in the function. When a particular outcome is realized through measurement, the wave function undergoes a sudden change—known as collapse—from a superposition of possibilities to a definite state corresponding to the observed property. This collapse is triggered by the interaction of the quantum system with an external observer or measuring device, such as a Geiger counter or bubble chamber. The process introduces a brief discontinuity and irreversibility into the wave function, rendering the Schrödinger equation inapplicable during the collapse but valid again afterward. As physicist Paul Davies aptly remarked, "reality is in the measurements, not in the electron."

A more academic evaluation reveals that Bohr and Heisenberg themselves never fully agreed on how quantum mechanics should be interpreted. Neither of them used the term Copenhagen interpretation—the label was introduced later to denote the shared features of their views. While the Copenhagen interpretation clearly embraces indeterminism over determinism, its stance on realism versus anti-realism remains open to debate.

Both Bohr and Heisenberg described quantum mechanics within a logical positivist framework. Bohr was especially concerned with the philosophical implications of quantum theory, as exemplified by his complementarity principle. He believed quantum theory offered a complete description of nature, albeit one that defied classical intuition. While classical mechanics and probability were better suited for everyday experience, Bohr never specified a clear boundary between the quantum and classical domains. He saw the distinction as philosophical or a matter of practical convenience, rather than a scientific necessity.

Over time, the Copenhagen interpretation has faced numerous criticisms. Key objections include the discontinuous "jump" in the wave function during measurement, the introduction of randomness into observation, the subjective role assigned to the observer, the ambiguity in defining what constitutes a measuring device, and the reliance on classical physics to describe the measuring environment.

De Broglie–Bohm Interpretation

Let us return to Smith once again, but this time to another of his important works, *The Wisdom of Ancient Cosmology*:

> Already in 1927, when the heavyweights of physics gathered in Brussels to assess what (if anything) was still left of the Newtonian *Weltanschauung*, Niels Bohr and Albert Einstein took up diametrically opposing positions on this question which have polarized the debate ever since. What Bohr said, in effect, is that when it comes to quantum phenomena, we must be satisfied with the pragmatic kind of information the quantum theory delivers. It is the best one can do, not because the theory as such is imperfect or incomplete (as Einstein charged), but because there is in fact no "deep quantum reality" behind the phenomena

which could explain what is going on. "There is no quantum world," said Bohr. "There is only an abstract quantum description."

Einstein, on the other hand, argued with all his considerable might in favor of a realist interpretation. There is a "quantum world," he said in effect, a deep reality which underlies the phenomena and ultimately explains them: there must be! It was unthinkable to him that an "abstract quantum description," unsupported by any deep reality, could lead to correct predictions as if by magic.

As it turned out, most physicists eventually joined the Bohr school of thought, which came to be known as the Copenhagen interpretation of quantum theory. In 1932, moreover, just five years after the Brussels conference, it seemed very much as though the Copenhagenist claims had been vindicated, once and for all, through the labors of a Hungarian mathematician by the name of John von Neumann. What von Neumann had done, first of all, was to put the new quantum theory on a firm mathematical foundation, and then prove, with exemplary rigor, that the idea of a physical reality consisting of "ordinary objects" is mathematically impossible (that is to say, incompatible with the predictions of quantum theory). At this point it appeared to all but a few realist diehards—led by the unconquerable Albert Einstein—that the issue was henceforth closed.

But in fact it was not. The realist position revived, phoenix-like, in 1952, when David Bohm—after extensive conversations with Einstein—succeeded in constructing an objective model of the electron which squared with the exacting demands of quantum theory. In the wake of the celebrated "von Neumann proof," Bohm had apparently accomplished the impossible. It thus became clear that there must be a loophole in von Neumann's argument; and yet the great mathematician had done his work so well that it took another twelve years to find it! The riddle was not resolved until 1964, when a hitherto unknown physicist by the name of John Stewart Bell discovered that von Neumann had unwittingly made one hidden and unwarranted assumption concerning the "ordinary objects" which quantum theory supposedly ruled out. What von Neumann had assumed—and what until then everyone apparently regarded as self-evident that the objects in question were "local" entities. Basically this means that they can only communicate with each other via known physical forces: via signals, therefore, which cannot propagate faster than the speed of light. It turns out, moreover, that Bohm was able to get around the von Neumann interdict precisely because his model of the electron did not in fact obey the stipulated locality condition.

But this result, this decisive breakthrough, proved to be only the beginning. Basing himself on the fundamental principles of quantum theory, Bell succeeded in proving that reality—be it ordinary or contextual!—must in fact violate von Neumann's locality condition. What Bell proved finally is that the locality condition is not only unwarranted, but actually untenable. In a word, reality is nonlocal. This is the amazing discovery which has come to be known as Bell's theorem. It has since been verified experimentally on the basis of an inequality (likewise discovered by Bell) which however does not hinge upon quantum theory; the result, therefore, appears to stand solid as a rock. But whereas most physicists today accept Bell's theorem as well-founded, a few continue to be skeptical and have left no stone unturned in an effort to thwart Bell's conclusion. Such exceptional scrutiny, moreover, is not unwarranted, considering the momentous implications of Bell's claim. Rightly understood, the affirmation of nonlocality revolutionizes our worldview. It is not without reason that Berkeley physicist Henry Stapp refers to Bell's theorem as "the most profound discovery of science."[3]

[3] *The Wisdom of Ancient Cosmology: Contemporary Science in Light of Tradition* (Oakton, VA: The Foundation for Traditional Studies, 2003).

In the De Broglie–Bohm theory (also called Bohmian mechanics), particles have definite positions at all times. These positions evolve deterministically under the guidance of the wave function, which itself evolves according to the Schrödinger equation. Notably, in this interpretation, the wave function does not collapse upon measurement; instead, it becomes effectively confined to a particular component of the wave, giving the appearance of collapse.

This approach treats the wave function as a real, physical field—a stark contrast to the Copenhagen interpretation, where it is merely a mathematical abstraction. In De Broglie–Bohm theory, quantum mechanical entities have ontological primacy over classical concepts, meaning they reflect the fundamental nature of reality rather than mere tools for prediction.

Because the theory posits hidden variables—such as definite particle positions—it is often described as a hidden variable theory. It is deterministic (the future is completely determined by the present), and nonlocal (it allows instantaneous connections between distant particles), thus satisfying the requirements of Bell's theorem.

Although the theory restores realism, it still conforms to the Heisenberg uncertainty principle: while particles possess definite positions and momenta, our ability to measure both simultaneously remains limited by quantum constraints.

Ultimately, the De Broglie–Bohm theory offers a coherent and compelling alternative to the Copenhagen interpretation—one that dares to take the wave function seriously as part of physical reality and challenges us to rethink the very fabric of the universe.

Multiverse Interpretation

The Many-Worlds (or Many-Universe) interpretation of quantum mechanics posits the existence of a universal wave function as an objective reality. According to this view, the wave function never collapses; instead, it evolves continuously under deterministic and reversible laws. Since Hugh Everett's (1930–1982) seminal 1957 paper, numerous variants of this interpretation have emerged.

In this framework, the phenomenon observed during measurement is explained through decoherence. Decoherence refers to the suppression of interference (i.e., quantum superposition) in the wave function due to the system's interaction with its environment. Here, "environment" refers not to classical objects like slits or screens (as in the double-slit experiment with electrons), but to external quantum systems. For instance, when a sufficiently large number of cosmic particles randomly strike an electron traveling between the slits and the screen, the resulting interaction leads to entanglement between the electron and its environment. This entanglement suppresses the interference pattern in the wave function, giving rise to classical behavior—a process referred to as decoherence. In this way, decoherence bridges the gap between quantum formalism and classical experience, thus addressing the measurement problem.

The Multiverse approach resolves the apparent randomness of the Copenhagen interpretation by asserting that all possible outcomes encoded in the universal wave function are realized—each in a distinct universe. When we open the box to observe Schrödinger's cat, for example, or perform any measurement, reality "branches" into multiple universes: in one, the cat lives; in another, it dies. Each of these resulting worlds represents a distinct, equally real state of the universal wave function. These worlds are non-interacting. Since the universal wave function encompasses all of these realities, it may be more accurate to call it a multiverse wave function. This wave function does not collapse; instead, we merely find ourselves in a universe that corresponds to one particular outcome. Unlike single-universe interpretations that present a linear history of reality, the multiverse approach envisions history as a branching tree of almost infinite possible quantum events. Consequently, there may be countless versions of ourselves inhabiting parallel universes—none of which are aware of each other.

A notable variant of this interpretation is the Many-Minds hypothesis. This view proposes that it is not the universe that branches, but rather the mind of the observer. The motivation behind this variant is to avoid the bizarre implication that an observer could be in a superposition of mental states. In their 1988 paper, David Albert and Barry Loewer argued that it is incoherent for an observer to consider their own mind as being in an indeterminate state—one must be certain about the state of the system being observed. If the mind itself were in a superposition, such certainty would be impossible, leading to a contradiction. To resolve this, they suggested that while the body (i.e., physical substrate) of the observer may exist in a superposed state, the mind remains in a definite, non-superposed state.

Interpretation of Collapse Induced by Consciousness

Also known as the von Neumann–Wigner interpretation, this approach—of which several variants exist—centers on the distinctive idea that consciousness is essential for completing the quantum measurement process. In his 1932 work, *Mathematical Foundations of Quantum Mechanics*, John von Neumann (1903–1957) argued that the formalism of quantum mechanics permits wave function collapse to occur at any point along the causal chain—ranging from the measuring apparatus to the observer's subjective perception of the measurement.

In the 1960s, Eugene Wigner (1902–1995) extended this idea by reformulating the well-known "Schrödinger's cat" thought experiment into the now-famous "Wigner's friend" scenario. Wigner proposed that, regardless of one's realist stance, the true boundary where wave function collapse occurs is the consciousness of the observer. Recently, an extended version of the Wigner's friend experiment was reportedly tested, with claims that it provided empirical support for the idea that the quantum realm is not entirely independent of the observer.

Physicist Rudolf Peierls (1907–1995) was also a supporter of this interpretation. Similarly, Zvi Schreiber (b. 1969), in his master's thesis, articulated the view that the only true measurement device is the non-physical mind. He summarized this position as follows:

> The rules of quantum mechanics are valid, but only one system—the entire material universe—can be treated quantum mechanically. External observers, such as the human (and possibly animal) mind, lie outside the framework of quantum mechanics. Measurements that trigger wave function collapse occur in the mind.[4]

Henry Stapp further developed and defended this view, emphasizing the necessity of consciousness in any consistent quantum framework:

> From the standpoint of quantum theory's mathematics, there is no reason to treat a measuring device as anything other than a composite of atomic components—it is simply another part of the physical universe... Moreover, the conscious experiences of the human observer must have a direct and immediate causal relationship with what is occurring in their own mind, not with events in an external apparatus... Therefore, our bodies and brains become part of the physical universe described by quantum mechanics. Treating the universe as an integrated whole offers a conceptually simple and logically coherent foundation.[5]

Despite these arguments, many physicists remain skeptical of this interpretation. They argue that the notion of consciousness causing collapse relies on interactive dualism, which conflicts with the materialist assumptions underpinning modern physics. This raises the unresolved question of how consciousness

[4] Schreiber, Zvi (1995) "The Nine Lives of Schrödinger's Cat," Imperial College London, M.Sc. dissertation.

[5] Stapp, Henry, "Quantum Theory and the Role of Mind in Nature," *Foundations of Physics*, 31 (10): 1465–99 (2001).

interacts with matter—an issue reminiscent of philosophical objections to Descartes' substance dualism.

Albert Einstein (1879–1955), a staunch scientific realist, acknowledged the epistemic utility of the wave function but rejected any notion that consciousness could affect the physical state of matter. He viewed quantum mechanics as an incomplete theory rather than a final description of reality, and did not accept the idea that subjective observation determines objective outcomes.

Consistent Histories Interpretation

The *Consistent Histories* approach—supported by physicists such as Murray Gell-Mann (1929–2019), Nobel laureate and developer of quark theory—offers a framework for understanding which sets of classical questions can be consistently applied to a single quantum system, and which sets are fundamentally incompatible and therefore meaningless when asked together. This interpretation provides a resolution to the paradox posed by Einstein, Podolsky, and Rosen (EPR), who assumed that multiple classical properties could be simultaneously well-defined for a single quantum entity.

The Consistent Histories framework clarifies why such simultaneous inquiries are invalid within quantum mechanics, while also explaining why classical logic often appears valid in many quantum experiments. Importantly, it reveals that there are mathematical boundaries within which classical reasoning holds, thereby providing a precise account of where and why classical logic breaks down in the quantum domain.

Statistical Interpretation (Statistical Community Perspective)

According to the *Statistical Interpretation*, the wave function does not describe individual quantum systems, such as a single particle, but rather represents an abstract statistical quantity applicable only to ensembles of identically prepared systems. This view was most famously advocated by Albert Einstein, who wrote:

> The attempt to regard the quantum-theoretical description as a complete representation of individual systems leads to strange theoretical consequences. But if one assumes that this description refers only to ensembles of systems, the difficulties vanish.

Proponents of this interpretation describe it as minimalist, making the fewest assumptions about the physical meaning of the standard quantum formalism. It takes Born's statistical interpretation to its logical conclusion, viewing quantum mechanics purely as a tool for calculating probabilities across ensembles.

More recently, an updated version of this approach has been proposed, grounded in a new formulation of probability theory. It has been shown that the Kolmogorov–Markov (KM) laws naturally emerge from this new probabilistic framework, reinforcing the statistical interpretation's mathematical consistency and predictive power.[6]

Relational Interpretation

The *Relational Interpretation* posits that the quantum state of a system is relative to the observer—that is, it reflects the relationship between the system and the observer, rather than the system's objective, standalone properties. First introduced in 1994 by Carlo Rovelli (b. 1956), the interpretation has since been further developed by various theorists. It draws inspiration from the observer-dependence of measurements in special relativity, and also incorporates ideas from John Wheeler's Quantum Information Theory.

According to Rovelli:

[6] Shaiia, Raed M., "On the Measurement Problem," *International Journal of Theoretical and Mathematical Physics*, 4 (5): 202–19 (2014).

> Quantum mechanics is a theory about the physical description of physical systems in relation to other physical systems, and this theory is a complete description of the world.[7]

At its core, relational quantum mechanics suggests that different observers may construct different, yet internally consistent accounts of the same physical events. For instance, a quantum system may appear to be in a definite eigenstate to one observer, while simultaneously appearing as a superposition to another. If quantum mechanics is indeed a complete theory, then the concept of "state" does not describe the system in isolation, but rather the correlation between system and observer.

Notably, the term *observer* here refers to any physical system, microscopic or macroscopic—it does not imply consciousness. In this view, a "measurement" is simply a physical interaction that establishes a correlation between systems.

This relational framework elegantly avoids certain paradoxes associated with nonlocality. Unlike interpretations that require faster-than-light information transfer (as seen in some readings of Bell test experiments), the relational interpretation maintains locality for all observers, as each observer's account is internally consistent and no absolute frame of reference is required.

Operational Interpretation

Physicist John G. Cramer (b. 1934) proposed the Operational Interpretation of quantum mechanics, articulating it in the abstract of his 1986 paper:

> The basic element of this interpretation is the transaction describing a quantum event as an exchange of advanced and retarded waves, as implied by the work of Wheeler and Feynman, Dirac, and others. The transactional interpretation is explicitly nonlocal and thereby consistent with recent tests of the Bell inequality, yet is relativistically invariant and fully

[7] Rovelli, Carlo, "Relational quantum mechanics," *International Journal of Theoretical Physics*, 35: 1637–78 (1996).

causal. A detailed comparison of the transactional and Copenhagen interpretations is made in the context of well-known quantum-mechanical *Gedankenexperimente* and "paradoxes." The transactional interpretation permits quantum-mechanical wave functions to be interpreted as real waves physically present in space rather than as "mathematical representations of knowledge" as in the Copenhagen interpretation. The transactional interpretation is shown to provide insight into the complex character of the quantum-mechanical state vector and the mechanism associated with its "collapse." It also leads in a natural way to justification of the Heisenberg uncertainty principle and the Born probability law ($P = \psi\psi^*$), basic elements of the Copenhagen interpretation.[8]

Stochastic Interpretation

The Stochastic Interpretation presents a classical reinterpretation of the Schrödinger equation, drawing analogies with Brownian motion. This approach was formally proposed by Edward Nelson (1932–2014) in 1966, although similar ideas had been published earlier and were acknowledged in Nelson's work.

In this interpretation, the quantum behavior of particles arises from an underlying stochastic (i.e., both random and deterministic) process. The core idea is that the apparent quantum fluctuations of space or spacetime are not a *consequence* of quantum mechanics, as traditionally thought, but rather its *cause*. Recently, new formulations of the stochastic interpretation have emerged, continuing to explore this classical yet probabilistic foundation for quantum theory.

Objective Collapse Interpretation

The Objective Collapse Interpretation diverges from the Copenhagen view by asserting that both the wave function and its

[8] Cramer, John, "Transactional Interpretation of Quantum Mechanics," *Reviews of Modern Physics*, 58 (3): 795–98 (1986).

collapse are objectively real phenomena—not merely tools for predicting observations. According to this interpretation, wave function reduction is a physical process that occurs either spontaneously (randomly) or when a specific physical threshold is reached. The observer plays no essential role.

These theories are realist and indeterministic, rejecting the need for hidden variables. Unlike standard quantum mechanics, which provides no explicit mechanism for collapse, objective collapse theories aim to extend the formalism to include one. Notable examples include the Ghirardi–Rimini–Weber (GRW) theory, Roger Penrose's interpretation, and deterministic variants within this class. If such theories are correct, quantum mechanics must be reformulated not merely as a predictive framework, but as a complete physical theory that explicitly accounts for collapse.

Quantum Logic

Quantum logic is a reformulation of propositional logic designed to address the apparent contradictions between classical Boolean logic and quantum measurement. Introduced in a 1936 paper by Garrett Birkhoff and John von Neumann, it seeks to reconcile the peculiar behavior of complementary variables and the outcomes of quantum observations with a consistent logical framework.[9] Rather than modifying the physical laws, quantum logic alters the structure of logical propositions to better align with quantum phenomena, especially regarding measurement compatibility and the nature of quantum truth values.

Quantum Information Theories

Quantum Information Theories explore how information is encoded in quantum systems. While classical information theory

[9] Birkhoff, Garrett and von Neumann, John, "The Logic of Quantum Mechanics," *Annals of Mathematics*, 2nd Ser. 37 (4): 823–43 (1936).

uses bits as its basic unit, quantum information theory uses qubits, which can exist in superpositions of states. These theories are typically categorized into two broad philosophical camps:

Ontological Approaches: These include doctrines like John Wheeler's "It from Bit," which propose that reality is fundamentally composed of information. This view forms the basis of digital physics, where the universe is conceptualized as the output of a vast computational process—whether deterministic or probabilistic. In this framework, physical systems are essentially information structures, making such theories a modern form of immaterialism, the philosophical stance that denies the existence of a purely material world.

Epistemic Approaches: In contrast, these interpretations assert that quantum mechanics describes not objective reality, but an observer's knowledge or information about it. Collapse, in this context, is not an ontological event but an update of the observer's information upon measurement. These views often echo the philosophy of Niels Bohr, emphasizing instrumentalism, the idea that scientific theories are tools for organizing observations rather than literal descriptions of reality.

Model Interpretations of Quantum Mechanics

The original "Model Interpretation" of non-relativistic quantum theory, introduced by Bas van Fraassen (b. 1941) in the early 1970s, initially referred to a specific interpretation. Today, however, the term has broadened to encompass a category of interpretations that share certain foundational principles.

To address the measurement problem, van Fraassen proposed a distinction between two types of system states: the *dynamical state* and the *value state*. The *dynamical state* describes what a system *can* be—its potential physical properties at a given moment and in the future. The *value state*, by contrast, specifies what the system *actually is* at that moment—the definite physical properties it currently possesses.

In standard quantum mechanics, the dynamical state corresponds to the state vector, which evolves deterministically according to the Schrödinger equation and never undergoes collapse. What sets the Model Interpretation apart is the idea that even if a system's dynamical state is not an eigenstate of a particular observable, the value state can still assign that observable a definite value.

This perspective breaks with the conventional "eigenstate-eigenvalue link," which holds that a system can possess a definite value for an observable only when it is in the corresponding eigenstate. In Model Interpretations, the link is conditional: *if* a system is in an eigenstate, then the observable has a definite value—but the reverse is not necessarily true.

Time-Symmetric Theories

In *time-symmetric* theories, such as the Wheeler-Feynman absorber theory, quantum mechanical equations are modified to remain symmetric under time reversal. This implies a form of *retrocausality*, where not only do past events influence the future, but future events also influence the past.

In such models, a single measurement is insufficient to fully determine the state of a quantum system. However, if two measurements are made at different times, the complete state of the system between those times can be calculated precisely. As a result, the collapse of the wave function is reinterpreted not as a physical transformation, but as an update in our knowledge of the system—triggered by the second measurement.

Entanglement, from this perspective, is not a real physical connection between particles but an illusion arising from ignoring retrocausal effects. The apparent "entanglement" is the result of each particle being influenced by a future event involving the other.

While not all proponents of time-symmetric theories advocate modifying the unitary dynamics of standard quantum mechanics, notable figures like Lev Vaidman (b. 1955), who supports the *two-state vector formalism*, emphasize its compatibility with Everett's Many-Worlds Interpretation.

Bifurcating Spacetime Theories

Bifurcating Spacetime Theories resemble the Many-Worlds (or Multiverse) interpretation, but with a key distinction: bifurcation occurs not in the quantum state vector, but in the topology of spacetime itself. In the Many-Worlds view, it is the wave function that branches. In contrast, Bifurcating Spacetime theories posit that the actual fabric of spacetime undergoes bifurcation based on causal relations between events.

These interpretations have implications for Bell's theorem, quantum computation, and quantum gravity. They share conceptual ground with *hidden variable* theories and the *Statistical Community Interpretation*, which assumes well-defined particle trajectories at the microscopic level. However, at macroscopic scales, these trajectories can only be described stochastically—aligning with the probabilistic outcomes predicted by standard quantum mechanics.

The Silence Option

Although discussions on the interpretation of quantum mechanics are widespread today, this was not always the norm. Some prominent physicists have chosen to remain silent on interpretational matters. Paul Dirac (1902–1984), one of the pioneers of quantum theory, famously remarked:

> The question of how to interpret quantum mechanics has been considered by many authors, and I do not want to discuss it here. I want to deal with more fundamental things.[10]

[10] Dirac, P.A.M., "The inadequacies of quantum field theory," eds. B.N. Kursunoglu and E.P. Wigner, *Paul Adrien Maurice Dirac* (Cambridge: Cambridge University Press, 1987), 194.

This "silence option" remains a valid stance among many working physicists, who prefer to focus on the predictive success and practical applications of the theory rather than metaphysical debates. Figures such as Nico van Kampen (1921–2013) and Willis Lamb (1913–2008) have openly criticized what they saw as speculative or unorthodox interpretations of quantum mechanics.

Thomist Interpretation of Quantum Mechanics

The Thomist Interpretation is a philosophical approach to the measurement problem in quantum mechanics, rooted in the metaphysical framework of St Thomas Aquinas (1225–1274). Unlike many interpretations that propose new mathematical formulations, the Thomist Interpretation seeks to address the foundational puzzles of quantum theory using classical metaphysics—particularly the distinction between potentiality and actuality.

At its core, this interpretation posits that reality consists of at least two distinct ontological levels: the physical and the corporeal. Quantum mechanics—including the wave function and its formalism—pertains to the physical realm, which consists of potential beings that lack substantial form. In contrast, the classical appearances of objects, their definite properties, and the outcomes we observe in experiments belong to the corporeal realm, the realm of actuality and substantial form.

According to this view, the collapse of the wave function—a central issue in the measurement problem—is not merely a mathematical or epistemological event, but a real ontological transition from the potential mode of being (physical realm) to the actual mode of being (corporeal realm). This ontological discontinuity is reflected in the breakdown of the linear evolution described by the Schrödinger equation at the point of measurement.

Wolfgang Smith draws an important distinction between his Thomist view and that of Werner Heisenberg, who famously interpreted the wave function as a "probability wave," likening it to Aristotle's notion of *potentia*. At first glance, Heisenberg's "world of *potentiae*" appears to match what Smith calls the microworld (physical realm), and his "world of beings and events" seems to align with the corporeal world. However, Smith argues that Heisenberg's distinction is ultimately one of scale rather than ontology. In Heisenberg's philosophy, macroscopic actuality emerges through complexity and aggregation of microscopic potentialities—a view that Smith finds inadequate.

In contrast, Smith insists that the transition from potentiality to actuality cannot be explained by scale alone. Rather, it occurs whenever we move from a subcorporeal (SX) to a corporeal (X) object, even at the macroscopic level. The fact that SX can be described using classical physics or reduced to atoms does not mean that X, the corporeal object, is also reducible. They exist on different ontological planes. SX exists in potentiality; X exists in actuality. Therefore, every part of the macroscopic objects of classical physics is as ontologically "potential" as atoms and subatomic particles, until it is informed by a substantial form and enters the corporeal realm.

This view represents a downward ontology, in contrast to the upward ontologies that dominate modern interpretations of physics. In most interpretations—whether Copenhagen, many-worlds, or decoherence-based—the flow of explanation proceeds from micro to macro: classical reality is seen as emergent from quantum processes through aggregation, decoherence, or information transfer. The Thomist Interpretation inverts this explanatory direction: reality descends from a higher ontological level (*actus essendi*), and the physical realm of quantum mechanics exists as a lower, less fully real stratum—a realm of potentiality awaiting actualization.

Smith's model also critiques the Cartesian bifurcation inherent in modern scientific worldviews, where a rigid split between mind and matter underlies both classical and quantum physics. The Thomist Interpretation rejects this dualism, proposing instead that even the descriptions of classical physics pertain to the physical (potential) realm, while the actual objects of perception—what we know and interact with—belong to the corporeal realm.

The corporeal world is characterized by substantial forms, which are absent in the physical realm. These substantial forms, known in Islamic philosophy as A'yān-i Sābite (أعيان ثابتة)—immutable archetypes existing in divine knowledge—are what give corporeal objects their actuality. In this framework, quantum particles do not confer existence upon things, but rather receive existence from these substantial forms, which precede and ground physical phenomena.

As Smith concludes,

> This ontological position is not merely a matter of philosophical speculation, but a position that has been imposed on us by physical discoveries and especially the results of quantum theory—provided that we adhere to a realist stance, of course.[11]

Ali Sebetci

[11] Smith, *TQE*.

Chapter VIII

Newtonian and Aristotelian Physics: *Wolfgang Smith's Path to Reconciliation*[1]

The major milestone of a truly mathematical account of motion was accomplished with the publication of *Principia Mathematica* by Sir Isaac Newton in 1687. Prior to Newton's groundbreaking treatise man's understanding of kinematics was largely ontological, with renowned figures in pre-Newtonian physics, such as Aristotle, positing teleological explanations for the motions of bodies.[2] Newton departed decisively from this paradigm by mechanizing the physics of motion and removing any semblance of teleology. To this end he proposed a revolutionary model wherein formless lumps of matter are governed by universal forces. From one perspective, Newton's mechanics took the first step towards the instrumentalized approach to physics prevalent today; this approach is characterized by ontological minimization and heavy reliance on operational concepts for empirical success. From another viewpoint, however, Newton's mechanics was anything but instrumental: for while Newton did eliminate metaphysics from prediction-making, his mechanics was still entrenched in objective metaphysical entities as the ultimate explanations of empirical results.

One intriguing example of this metaphysical dependence is Newton's concept of absolute space. According to Newton, there are two forms of space: "relational spaces," defined by

[1] See Aristotle, *Physics*, Book IV, in *Complete Works of Aristotle, Vol. 1: Revised Oxford Translation*, ed. Jonathan Barnes (Princeton University Press, 1984).

[2] Ibid.

measurable relations between objects (e.g., measurable distances and relative velocities); and "absolute space," a deeper, fundamental entity theorized to exist independently of the objects it contains, and the ultimate stage for all motion.[3] Crucially, absolute space functions solely as a metaphysical postulate in Newton's mechanics, a recognition that became the flashpoint for one of history's most intense scientific debates, between Newton and Gottfried Leibniz.

The Newton–Leibniz debate centered on whether space should be understood as a set of relations among bodies (relationism), or as an absolute, distinct from its bodies and their relations (substantivalism). Initially, Newton's position prevailed, supported by thought experiments like "Newton's bucket." Gradually modern consensus has shifted towards relationism, particularly with the introduction of Mach's principle and Einstein's relativity. Henceforth Newton's concept of absolute space would be largely regarded as an antiquated notion to be consigned to the history books.

In this essay I will attempt to show that Newton's concept of absolute space is not so problematic as many suspect; the conflict between relationists and substantivalists can be largely reconciled through adopting certain ontological distinctions proposed by mathematician and philosopher of physics Wolfgang Smith; from this perspective I will argue that absolute space and relationism are but two sides of the same coin. I will then use this conclusion to contend that certain issues in mechanics, traditionally viewed as physical, are actually metaphysical, and can be accounted for accordingly; these issues relate to the nature of reference frames and of physics as such. Finally, I will explore how these insights open the possibility of revitalizing Aristotelian mechanics, albeit in a qualified sense.

[3] *The Principia: Mathematical Principles of Natural Philosophy: A New Translation*, trans. I.B. Cohen and A.W. Berkeley (University of California Press, 1999), Scholium.

Wolfgang Smith's Ontology

Wolfgang Smith posits that Nature comprises two distinct realms. The first is the "corporeal world," which encompasses those bodies and sensory qualities we encounter with our ordinary five senses. Modern metaphysics tends to subjectivize such qualities; à la Locke, sensory or "secondary" qualities like color are often interpreted as an artifice of the mind rather than an inherent property of the external world. Wolfgang Smith vehemently opposes this notion, arguing that all sensorily perceived attributes are integral components of external objects, which he calls "corporeal." Henceforth I will refer to the mind-independent realm containing these attributes as the corporeal world.

The second realm in Smith's ontology of Nature is the "physical universe," which encompasses the measurable objects accessible to physics.[4] Smith further divides the physical into two subdomains. The first is the "transcorporeal" realm, which includes measurable objects not directly associated with a specific corporeal entity[5]; in a recent work Smith identifies transcorporeal objects directly with quantum systems.[6] The other subdomain, the "subcorporeal," consists of corporeal objects *as understood in their quantitative dimensions.*[7] For any corporeal object X, that is to say, there is an associated subcorporeal object SX. Unlike X, SX lacks sensory qualities and consists of the quantitative remains of X. Put differently, SX is X conceived in the measurable terms of physics—i.e., geometrically, thermodynamically, spatiotemporally, etc.[8]

Although SX is quantitative, it nevertheless participates in

[4] Ibid., 29.

[5] Ibid., 35.

[6] *Physics: A Science in Quest of an Ontology* (PSIF, 2023), 28–33.

[7] *The Quantum Enigma*, 34. A physical object—be it subcorporeal or transcorporeal—is, in the words of Smith, the corporeal object "as conceived by the physicist."

[8] Ibid.

the reality of X, leading Smith to his next core distinction between wholeness and irreducible wholeness (IW). In Smith's ontological framework, irreducible wholeness transcends a mere summation of parts, contrasting with ordinary wholeness which is simply a collection of parts.

A classic example of an IW is a circle, since any instantiation of a circle by definition cannot be reduced to a mere collection of points on a graph. To illustrate this, imagine a graph filled with a random array of unconnected points scattered across the entire plane. Despite the possibility of connecting these points to form a circle, no rational observer would perceive a circle within this chaotic arrangement before doing so. This indicates that a circle is not reducible to a set of points on a graph and thus constitutes an IW. Furthermore, even if we imagine a collection of points arranged in a circle, it is clear that the concept of "circle" must precede the arrangement of these points; otherwise, it would be impossible to organize them into this shape in the first place. This fact was recognized by the famous Gestalt psychologist Wolfgang Köhler,[9] who reiterated Aristotle's insight that the whole is more than the sum of its parts.

Smith asserts that in the corporeal world any object X is necessarily an IW, a feature which is further reinforced by the possession of what St Thomas Aquinas called a "substantial form," which arranges matter into one thing rather than another. And since all corporeal objects X are irreducible wholes, their counterpart objects SX are as well. However, the irreducible wholeness of SX is purely quantitative and constitutes the objective classical—physical properties also found in its associated corporeal object X, via participation of the former in the being of the latter.[10]

[9] *Gestalt Psychology: An Introduction to New Concepts in Modern Psychology* (Liveright, 1947).

[10] *Physics: A Science in Quest of an Ontology*, 26–27.

Substantivalism vs. Relationism

In light of Wolfgang Smith's ontology, we can now turn to the challenge of reconciling substantivalism and relationism. In essence my proposal is that relationism pertains to the subcorporeal domain. As we ascend to the corporeal, however, we encounter more than just spatial relations: we find an absolute space that exists independently of the bodies which it contains.

When casting our gaze onto the corporeal world we encounter a space that transcends measurement. We perceive an independent plenum distinct from its contained bodies and the measurable relations between them. Space, in other words, appears to be more than a sum of parts and to possess a kind of corporeality or perceptibility. This is one reason why I propose that corporeal space should be considered substantival, exceeding the sum of its measurable parts and relations. In fact, within the corporeal realm, space assumes an identity of its own, as an IW, and as an independent container-boundary,[11] owing to its perceptibility.[12]

When we descend from the corporeal to the subcorporeal, what remains are only the measurable relations between objects, such as distances. And this aligns with the fundamental approach of physics, which concentrates on such relations without invoking the notion of absolute space. Wolfgang Smith's distinction between the corporeal and the physical provides the basis of reconciliation between substantivalism

[11] By "container-boundary" I mean space conceived as a cosmic "bound." However, as Smith points out, space as a bound is not incompatible with it also being an "empty container"; i.e., a distinct substance containing objects. Furthermore, it could be that the boundary aspect of space takes precedence, while the container aspect depends parasitically on the former.

[12] Independent perceptibility means that corporeal space is perceived as an entity distinct from its contained bodies. This does not entail that other corporeal bodies may not be required to aid in our perception of such a space, e.g., with visual perception that might require objects as reference points to gauge the presence of an independent space.

and relationism, allowing for the validity of both perspectives, albeit at different ontological levels.

While the corporeal domain comes with a space that transcends measurable relations, it is nonetheless true that corporeal space is measurable. The reason for this is that although space can indeed be bifurcated into a measurable collection of relations on the one hand, and something greater on the other, the measurable conception of space at the subcorporeal still participates in non-measurable, corporeal space—analogous to how a subcorporeal object SX participates in a corporeal object X, whilst also maintaining an identity distinct from X.[13] Therefore, although substantivalism and relationism are both valid notions in their respective domains, there is still a sense in which the relational conception of space is ancillary to the substantival one; namely, the former is the measurable dimension of the latter.

The considerations here about space also apply to the nature of time, with one major difference. For Wolfgang Smith, time does not originate in the corporeal world, but in a higher domain which he calls the "intermediary." The intermediary level consists of an uninterrupted flow of cosmic time, unbound by space, with no bodily parameters or relations. This cosmic time is essentially immeasurable, but also determines the measurable temporal relations from the corporeal level downwards. To use the lexicon of modern physics, "substantivalism" with respect to time is valid at the intermediary level, and "relationism" with respect to time is valid at the corporeal and subcorporeal levels.

Once more, we can also surmise that the measurable temporal relations between corporeal bodies participate in the immeasurable time of the intermediary. This is so because cosmic time determines and sustains these measurable relations. However, the measurable temporal relations between bodies are not essential to time as such; time would still be conceivable

[13] *Physics: A Science in Quest of an Ontology*, 27.

without bodies in measurable temporal relations, just as space is conceivable substantively without measurable spatial relations. Instead, these measurable relations derive their existence through participation in something immeasurable. As Plato remarks, "if the one has no participation in time whatsoever, it neither has become nor became nor was in the past"![14]

Teleology, Motion, and Physics

The foregoing analysis sets the stage for what concerns me in this section: the relationship between teleology, frames of reference, and physics.

The principle that "all inertial frames of reference are equal" is well known to any physics student. This principle essentially asserts that between any two non-accelerating reference frames, neither can claim an absolute state of motion over the other. In other words, no inertial frame can rightfully claim to be really stationary while deeming the other as really moving.[15] Claims about motion must rather be relativized to the perspective of specific inertial frames of reference. However, following the previous section, the existence of absolute space suggests that this principle may be incomplete. The existence and corporeality of absolute space implies that inertial motion extends beyond being a merely physical phenomenon and so in a sense must be absolute.

In this context I will contend that certain inertial reference frames can indeed be privileged over others, though on a non-physical, or corporeal, basis.[16] I propose that this privilege is teleological. After arguing for and outlining said proposal I will

[14] *Parmenides*, 141e.

[15] Tim Maudlin, *Philosophy of Physics: Space and Time* (Princeton University Press, 2012).

[16] This is not to dismiss the possibility of physically privileging certain frames of reference. However, the true basis of such privilege is fundamentally non-physical, with any physical manifestation merely reflecting this deeper underlying principle as a quantitative signature.

close this section by considering its ramifications for our broader understanding of the relationship between physics and motion.

Consider the textbook example of a train moving along some tracks. According to physics, it is entirely coherent to argue that the tracks are moving while the train remains stationary.[17] And in a certain sense this assertion holds true because what the physicist perceives is not the train per se, but the train's associated physical object SX. When we literally open our eyes we of course do not see SX but a corporeal object X, with its own distinctive nature and substantial form—i.e., a train. Moreover, on the corporeal level, the train discloses itself as inherently designed for motion—its *telos*, or end, is directed towards movement, whereas the tracks have a different end. From this we can infer that it is the train, and not the tracks, that is truly in motion. The commonsense judgement (or perspective) of a bystander—that the train is moving rather than the tracks—ultimately proves to be correct.

To flesh this out a bit, consider the actual engineering of a moving steam train. The train harnesses steam power, generated by water in its boiler, which then drives pistons that connect to the wheels, propelling the train forward. This process entails an intrinsic orientation towards motion on the part of the train. We therefore have valid reason to privilege the train's reference frame over that of the tracks. The train moves while the tracks remain stationary when in relative motion. Granted, this conclusion is only reached on a non-physical basis and primarily at the corporeal level. Nor does this perspective have any bearing on physics as such, which views inertial systems as measurable aggregates with equal claims to privileged states of motion—at least in terms of raw kinematics.

[17] Mathematically, from the perspective of any inertial frame, that frame is always treated as being at rest when constructing spatiotemporal coordinates. This perspective is crucial, which is why, from the viewpoint of someone on an inertially moving train, the train is perceived as stationary while the tracks appear to be moving.

Obviously this analysis is applicable to any system intrinsically ordered towards motion, from turtles to spaceships. In general, I propose that when two objects are moving relative to each other inertially, the one teleologically predisposed towards movement via its corporeal manifestation should be privileged as being "in motion" over one that does not.[18] This privilege, in turn, extends to the object's associated physical system SX.

But how should we interpret situations where there is no obvious disposition toward motion at all? For example, a rock moving inertially after being thrown, or falling toward the earth under the influence of gravity at terminal velocity?

Such scenarios typically fall into two categories as outlined by Aristotle in his *Physics*: (1) motions that occur naturally or stem from the inherent locomotive tendencies of bodies; and (2) motions that occur violently or go against the natural locomotive tendencies of bodies.[19] The case of a rock falling towards the earth constitutes an example of "natural motion." This movement results from the rock's inherent inclination to move towards the center of the Earth, where it finds its "natural place of rest."[20] In contrast, a rock being thrown constitutes a case of violent motion, as this goes against the rock's natural tendency to be at rest near the Earth's center.

In instances of natural motion, there are inherent dispositions toward movement, though they are not readily apparent. To fully discern these tendencies, I propose that one must consider them within some broader metaphysical framework of motion, as seen in Aristotle's physics for example. When analyzing two reference frames in relative inertial motion, priority should be given to

[18] Provided, of course, that the object is judged as being in motion by the opposite frame. This principle only applies to cases where one object bears a *telos* towards motion and the other does not. In scenarios involving multiple objects with an intrinsic orientation towards motion, a new principle must be established—perhaps deriving from the teleological context of certain actions in the corporeal world.

[19] See Aristotle, *Physics*, Books IV and VIII, op. cit.

[20] Ibid., Book IV.

the frame in which the corporeal object exhibits a predisposition towards motion, rather than the frame in which the object does not. In instances of violent motion within the corporeal realm, this motion originates from one object A causally interacting with another object B. B either has an inherent end (*telos*) of being in motion itself or of imparting motion to A. In cases of violent inertial motion, the objects involved in this causal interaction should be regarded as "in motion" over objects that are not. Therefore, in cases where dispositions toward motion are not obvious, real motion at the corporeal level can still be discerned teleologically. Once again, and to reemphasize, these conclusions about motion do not change anything concerning physics per se, since they pertain to the corporeal domain.

The aforementioned recognitions invite us to reconsider the relationship between physics and motion. More specifically, they compel us to ask whether physics truly deals with motion properly so called, with real changes in spatial place over time.

As we have noted, physics—with the possible exception of electrodynamics—treats all inertial reference frames equally. This suggests that in terms of raw mechanics the physics of inertial systems is, in actual fact, blind to motion. If every system can claim to be in motion, it would seem that, in a more fundamental sense, none truly is. It seems there are no real changes in spatial place over time, because when each and every system can lay claim to being in motion at the expense of every other, they all effectively cancel each other out. This rings especially true in light of the fact that physics is blind to the teleological origin of motion—the factor which I have said distinguishes inertial frames of reference. Therefore, when it comes to inertial systems, physics does not deal with motion as such. It rather deals with a surrogate kind of motion wherein all systems have equal privilege and bodies "move" relative to each other.[21]

[21] For the reasons stated, relative motion turns out not to be a meaningful concept.

This surrogate, "physical" understanding of motion is intelligible to us only because it is anchored in our corporeal understanding of motion with its attendant teleological and ontological entailments. The physicist's understanding of inertial motion is in fact parasitic upon such motion as we observe in the corporeal world. The criteria for distinguishing between true and false motion are established at the corporeal level, analogous to how a building's foundation supports the structure that rests upon it. Just as understanding the superstructure's function does not occur in isolation, the same applies to understanding the physicist's conception of inertial frames.

When it comes to scenarios involving acceleration, it seems prima facie that physics does deal with motion. The acceleration of a body is, after all, agreed on universally by all inertial frames and so is considered an absolute in Newtonian mechanics. But this is also misleading: strictly speaking, the Newtonian law of motion which relates the acceleration of a mass particle to an external force acting upon it is primarily concerned not with motion as such but with its variation. And rightly so, given that acceleration is meant to represent some change in a pre-defined quantity (velocity) rather than provide a value for that quantity.

It would appear that physics qua physics can actually tell us nothing concerning rest and motion in any absolute sense—not because there is no such thing as rest or motion, but because physics concerns notions that either ground motion conceptually or deal with its variance. This blind spot in physics comes about as a result of the field's exclusion of absolute space and the teleology of inertial frames at the corporeal level.

Towards a Reintegration of Aristotelian Physics

We have discussed two pivotal propositions. The first is that there is such a thing as absolute space, but that its existence is corporeal. The second is that some inertial frames of reference can be objectively privileged over others on teleological grounds.

These recognitions prompt us to re-evaluate the relevance of Aristotelian physics—a paradigm which has long been considered obsolete—and I contend that an Aristotelian physics of bodies, together with the modern physicist's conception of motion, are two indispensable aspects of an even greater account of motion. Before presenting my thesis, however, I will first present a brief overview of Aristotelian physics and its relevance within the context of modern physics.

The basic principle underlying Aristotelian physics is that motion is a form of change from potency to act. For Aristotle there are thus two kinds of change or "motion": the kind seen when a substance alters in its accidental characteristics, and change in the spatial location of a body over time.[22] Both types of motion involve the "actualization of that which was once potential."[23] For the purposes of this essay I use the term "motion" to refer to the second kind of change, in keeping with how we have used the term hitherto.

In his account of motion Aristotle makes use of two concepts to explain this phenomenon. The first we have seen already: teleology. The second is the concept of force. For Aristotle, of course, force is not understood in the same sense as in modern physics; i.e., as an influence propagated through a locally contained and mathematically describable field (usually represented by scalars or vectors). Rather, for Aristotle force is best understood as a power emanating from the immeasurable substance of a body.[24] Despite its difference from modern physics, Aristotle's notion of force does nonetheless bring about change in spatial place over time—it is just that this change is not alleged to be mediated through mathematized fields, i.e., "physically." By employing the concepts of teleology and force, Aristotle constructed a theory of motion that applies across a wide range of contexts, most notably

22 Aristotle, *Physics*, Book III.
23 Ibid.
24 Ibid., Book VIII.

celestial and terrestrial motion (only the latter concerns us here).

As we noted earlier, one of Aristotle's key distinctions is that between natural and violent motion. Natural motion arises from an object's inherent end (*telos*), driving it toward its proper place on Earth,[25] while violent motion results from some disruption to an object's natural motion by a force from another body.[26] The natural motions of objects for Aristotle are governed by numerous principles which relate to the five elements; he held that the element of earth naturally tends to rest at the center of the Earth, while the elements of water, air, and fire tend to ascend away from the center, each to different extents.[27] These behaviors reflect a teleological orientation possessed by the elements and the objects they compose. In contrast, violent motions occur when an external force is applied to an object, disrupting natural movement; throwing a rock into the air, for instance, represents a case of violent motion, as it interrupts the rock's natural inclination to move toward the earth. Aristotle also argued that force is transmitted through "contact" between two bodies.

These foundational ideas led Aristotle to conclusions often considered at odds with modern physics; most notably, that the natural motions of fire, air, water, and earth result from a teleological drive toward their destined places. Today we interpret such phenomena through the lens of the principles and forces of modern physics; we say the rock falls to the ground due to Earth's gravitational field, not because it harbors an intrinsic propensity to settle at the planet's core. Similarly, air rises because of its lower density, not because it "desires" to ascend to a particular place. Moreover, unlike Aristotle's metaphysical explanations, modern physics brings the advantage of being mathematically grounded, allowing precise and testable predictions about the natural world.

[25] Ibid., Book IV.
[26] Ibid., Book VIII.
[27] Ibid., Book IV.

Another way that Aristotle's physics differs from modern physics is that, in the case of violent motions, Aristotle posited that a force is always required to keep a body moving.[28] This obviously contradicts Newton's first law of motion/inertia, which states that an object at rest remains at rest, and an object in motion remains in motion—at constant speed and in a straight line—unless acted upon by an unbalanced force.

Such discrepancies have caused most to dismiss Aristotle wholesale. Carlo Rovelli, on the other hand, suggests we should not rush to judgment on this score. For instance, Rovelli remarks that Aristotelian physics can actually be seen as an approximation of Newtonian mechanics when applied to bodies moving through fluids.[29] The notion that Aristotelian physics simply contains no truth whatsoever is mistaken. And in light of Wolfgang Smith's account of reality, in particular, the physics of Aristotle appears to be quite relevant indeed. In fact, as I will now argue, the modern physicist's understanding of motion and Aristotle's are not only compatible but, when combined, afford us a more profound and illuminating view of motion.

First of all, in the consideration of whether natural motion is driven by forces and physical fields or by teleological orientation, there is no reason why these two explanations cannot coexist on different ontological strata, corresponding to Smith's physical and corporeal domains. At the corporeal level, teleological explanations account for the generalized motions of bodies. Meanwhile physical explanations involving forces and fields apply to motion at the physical or quantitative level. The two are complementary, insofar as teleological explanations address the broader question of why motions occur in the way they do, while physical explanations account for the measurable behaviors that arise as bodies fulfil said generalized motions. These allegedly "contradictory explanations" rather

[28] Ibid., Book VII.

[29] See "Aristotle's Physics: A Physicist's Look," *Journal of the American Philosophical Association*, Spring 2015, 1 (1): 23–40.

provide a more complete understanding of natural motion, from both qualitative and quantitative standpoints. Consider how the *telos* of a rock provides a generalized explanation for "why" it falls to the ground, while the quantitative notion of gravity—couched in the mathematical language of scientific theory—explains specified measurables displayed during the rock's free fall (such as velocity, gravitational acceleration, loss of potential energy, etc.).

Such complementarity between teleological and physical explanations indeed encompasses all manner of natural motion exhibited by the elements, as previously described. Teleological explanations, however, take precedence over physical ones, as they offer ultimate reasons for why motions occur. In this optic, physical explanations assume a secondary role, providing the quantitative characteristics of natural motion accessible by the *modus operandi* of physics.

Let us now consider the apparent conflict between Aristotle's views and Newton's first law of motion. This tension, I propose, may be resolved by acknowledging that Aristotle's concept of force is fundamentally different from Newton's. As previously noted, Aristotle conceived of forces as essentially powers or propensities effecting spatial change, emanating from the substantial forms of bodies. Crucially, he did not consider them to be measurable actions propagated through mathematized fields. This striking difference between Aristotle's conception of force and the modern scientist's—namely the fact that the two definitions are not in competition—opens the door to reconciling Aristotle's view on "eternal motion" with Newton's first law.

When it comes to forces as mathematized actions, Newton is absolutely correct in asserting that an object set in motion by a force will continue to move indefinitely in a vacuum without the continual application of that force (i.e., the first law). This fact is trivially true given the mathematical formalism of Newtonian physics. However, from the perspective of causal powers, Aristotle is also right in positing that the continual application of a force is needed to maintain motion: for if we view forces as emanating from the substance of bodies, as causal powers, there is no reason to think that forces do not continuously influence other bodies—even if they are not in direct physical contact! In fact, as a general principle, it can be understood that once an Aristotelian force causes a body to move or change spatially, it continues to exert influence "at a distance" on that body until another force acts upon it. Thus the Newtonian and Aristotelian perspectives on motion are complementary. Although it is accurate that a body can theoretically keep moving indefinitely in a vacuum, if it was initially set in motion by a Newtonian force from another body, its motion is perpetually maintained by an Aristotelian force operating ontologically between the two bodies, even in the absence of a Newtonian force.

These ideas presented above are admittedly in early stages and may seem incomplete at this point. However, with further development, these concepts could pave the way for a revival of Aristotelian physics. The initial step in this endeavor is to refine the integration of Aristotelian physics with classical mechanics at the terrestrial level. Following this, it will be important to explore how Aristotle's views on celestial bodies can be reconciled with modern, empirically supported understandings of the cosmos.

Closing Remarks

At the beginning of this essay, it was noted that the roots of modern physics can be largely traced to Sir Isaac Newton, whose revolutionary work laid the foundation for the field. Since then, however, physics has aimed to discard any notion of non-physical or absolute motion, which Newton himself advocated. As the arguments of this essay suggest, the endeavor to suppress such reckonings seems to have been ultimately misguided. In light of distinctions introduced by Wolfgang Smith, it is entirely reasonable to posit the existence of absolute space and time. We have also considered how Smith's ontological distinction between corporeal and physical reality allows for a harmonization between absolutist or substantivalist views of spatiotemporal reality and relationist perspectives. Moreover, we can now see that preferred and non-preferred frames of reference may be discerned on teleological grounds. All of which suggests that a radical reconception of the very nature of physics—exploiting the strengths of both mathematical and Aristotelian paradigms—is a very real possibility. What is needed next is a thoroughgoing deployment of the concepts and arguments advanced in this essay, foundations having been laid for an understanding of motion couched in Smith's ontology. Obviously, there remains much to be done in the application of these principles to more complicated examples and nuanced areas of analysis.

❋ ❋ ❋

May the conclusions drawn from this essay serve as a reminder of the rich and expansive nature of Wolfgang Smith's legacy—a legacy which I am sure has much more to offer mankind in his quest to understand reality. I strongly suspect that we have barely scratched the surface of what the categories and distinctions of Wolfgang Smith might reveal!

John Taylor

Chapter IX

Restoring an Integral Ontology

Wolfgang Smith on the Limitations of Scientism

Throughout Wolfgang Smith's extraordinary corpus of writings, he never failed to point out that "the philosophic foundations of physics are in disarray."[1] We may also reflect on the theoretical tenets of his work to examine the crisis facing modern psychology and mental health treatment today.[2] In this way, we can apply his penetrating appraisal of physics—and, modern science generally—to the discipline of psychology in the quest to rehabilitate its true ontology.

The imposition of a modern understanding of the human psyche, throughout most of the world (where the influence of the West is most pronounced) is nothing short of hegemonic and authoritarian. To date, this phenomenon remains unchecked. Psychology claims to be a science, yet falsely limits the scope of its knowledge. But what kind of knowledge does it adhere to? For there are levels of reality and distinct ways of knowing them. Certain epistemologies have access to transcendent horizons, while others are altogether confined to the five senses. All too often, we fail to ask if any kind of real wisdom is offered to us by psychology in its current state. The diverse cultures of traditional humanity have their own

[1] "Rediscovering the Corporeal World," *TQE*, 27.

[2] See Samuel Bendeck Sotillos, "The End of Quantum Reality: A Conversation with Wolfgang Smith," *Sacred Web: A Journal of Tradition and Modernity*, Vol. 45 (Summer 2020), 48–65.

(much more) profound insights regarding the human psyche, which are no less scientific, even if this word may not be used to describe them.

Smith elaborates as follows: "the one thing most needful for a just appraisal of modern science is the distinction between 'scientific knowledge' and 'scientistic belief'; that is to say, between *science* properly so called, and *scientism*."[3] The problem is that, in the present day, the two are confused and, in many ways, have become inseparable. It is often forgotten that the influence of human observation on scientific hypotheses is all too often overlooked. This has tremendous implications for empiricism. Smith writes that "Strictly speaking, there is no such thing as an 'empirical fact'—so long... as the term is understood to exclude the concomitant role of theory."[4] Hence, vigilance and discernment are required to distinguish true forms of science from mere ideology. He observes that "scientism is itself a kind of religion,"[5] adding that it "proves, in the end, to be the idolatry of a post-Christian civilization."[6]

A key reason why Smith's reflections on physics are important to other disciplines (including psychology) is the following: "[P]hysics proves ultimately not to be the basic science from which, in principle, all others are derived."[7] In this light, it is erroneous to restrict the object of all knowledge to corporeal reality alone. We tend to consider modern science as the sole arbiter of truth when, in fact, it is but one way of discerning truth among others. We need to remember, as he urges us, that "there are not only other modes of thought, but other modes of *bona fide* science as well."[8]

[3] "Modern Science and Guénonian Critique," *Science and Myth: What We Are Never Told* (San Rafael, CA: Sophia Perennis, 2010), 26.

[4] "What is the Physical Universe?," *TQE*, 42.

[5] "Science, Scientism, and Spirituality," *The Vertical Ascent*, 173.

[6] "Modern Science and Guénonian Critique," *Science and Myth*, 41.

[7] "Preface to the Second Edition," *Physics: A Science in Quest of an Ontology*, second edition, ix.

[8] Ibid.

The root of the crisis in modern psychology is the "Cartesian bifurcation"—the dualism between mind and body (along with matter) that has plagued the mindset of the West since the seventeenth century. It continues to be the default ontology of the modern West which, as needs to be made clear, has "no scientific basis."[9] In fact, Smith points out that "Cartesian bifurcation constitutes an ontological fallacy."[10] The outlook of Cartesianism was crucial for constructing the fractured metaphysics that pervades modernity today. Smith adds: "The Enlightenment has thus imposed an ontological inversion upon Western civilization: a rotation of 180 degrees if you will, interchanging the 'above' and the 'below.'"[11]

Smith speaks to the gravity of the situation facing modern science in this way: "There is... a price to be paid: for once the real has been split in two, no one apparently knows how to put the pieces back together again."[12] As a result of this myopic outlook, human beings are severed from reality and everything becomes objectified, further entrenching the psyche in a subject-object dichotomy. Cartesianists divide these substances into two aspects—subsequently dubbed as *res extensa* ("extended entities") and *res cogitans* ("thinking entities")—but fails to bridge the chasm that he thereby creates. Accordingly, all human experience is reduced to the private, subjective realm, thus eradicating any notion of objective reality.

This perpetuates the illusion of a fragmented world and an isolated self, which sunders our relationship to the sacred. This mind-body dualism lives on in modern science, especially in the fields of psychology and psychiatry, where this notion is deeply embedded in its epistemological framework. It is especially to be found in the *medical model* of clinical diagnosis and treatment of mental illness, which separates

[9] Ibid., 7.

[10] Ibid., 9.

[11] Ibid., 66.

[12] "Rediscovering the Corporeal World," *TQE*, 7.

the psychological (*psyche*) from the biological (*soma*). It is the metaphysical order that restores harmony to a consciousness that has been distorted by dualism. Deprived of an integral ontology, all scientific endeavors—including those in psychology—remain captive to relativism.

Smith expands on the implications of Cartesianism's breach, beyond just the mind-body split:

> One generally regards this Cartesian partition of reality into *res extensae* and *res cogitans* as simply an affirmation of the mind-body dichotomy, forgetting that it is much more than that; for not only has Descartes distinguished sharply between mind and body, but he has at the same time imposed an exceedingly strange and indeed problematic conception of corporeal nature, a conception, namely, that renders the external world unperceived and unperceivable.[13]

Although Smith discusses the human soul and its connection to the transpersonal order throughout his writings, he does not explicitly write about a "science of the soul." With this said, he presented his radical critique of modern psychology in two chapters of his book *Cosmos and Transcendence: Breaking Through the Barrier of Scientistic Belief* (1984); the first on Sigmund Freud (1856–1939) titled "The Ego and the Beast,"[14] and the second on Carl Jung (1875–1961), titled "The Deification of the Unconscious."[15] It is worth adding his reflections on modern psychology, which "is not only upside-down but dangerous in the extreme."[16] These chapters offer a powerful critique of how the discipline deviated from what it was originally intended to be, and how this process fueled the spiritual crisis of the modern world.

[13] "From Schrödinger's Cat to Thomistic Ontology," in *The Wisdom of Ancient Cosmology*, 41.

[14] "The Deification of the Unconscious," in *Cosmos and Transcendence*, 110–33.

[15] Ibid.

[16] "Astrology: The Science of Wholeness," *The Vertical Ascent*, 97.

The myriad fissures of the human psyche are direct consequences of the Enlightenment project, which was instrumental in catalyzing the loss of transcendence in modern science: "There can be no doubt about it: *the eclipse of transcendence constitutes the ultimate devaluation of the world*."[17] Because of this, the myriad crises all arising from the root cause, the spiritual crises of the modern world are inevitable, as Smith points out:

> In a society, namely, in which wholes are routinely reduced to a sum of parts, wholeness in its collective manifestations comes inevitably to be replaced by an abnormal and disruptive individualism lethal to all higher culture: one can see it happening nowadays in just about every cultural domain. The ongoing disintegration—beginning in the religious sphere, followed by the authentic arts—may thus be seen as the early effects of an ontological apostasy which neither the individual nor the collective can ultimately survive.[18]

It is worth adding that the desacralized cosmos of the Enlightenment has no *telos* or ultimate aim; it begins and ends in an ontological cul-de-sac. Smith affirms, "Cut off—as never before—from the source of our being, we have all but forgotten that life has meaning."[19] This inverted *Weltanschauung* unavoidably fuels nihilism, alienation, and disequilibrium: "The prevailing worldview, man is indeed a stranger in the universe, an accidental and ephemeral product of blind forces."[20]

All spiritual cultures of the world acknowledge that true unity is rooted in metaphysical discernment: "The great sapiential traditions, both of the East and the West, have emphatically declared... that wholeness precedes division."[21] This is a significant insight that points to the transpersonal dimension

17 "Appendix: Gnosticism Today," in *Teilhardism and the New Religion*, 240.

18 "Platonist Cosmology: The Long Lost Key," in *Physics: A Science in Quest of an Ontology*, 129–30.

19 "Science and Myth," in *Science and Myth*, 18.

20 "Introduction," *The Wisdom of Ancient Cosmology*, 13.

21 "The Tripartite Wholeness," *The Vertical Ascent*, 13.

of science when it is informed by an integrative ontology.

Smith thus introduces the concept of *irreducible wholeness* "which does not reduce to a sum of parts"[22] and where "a part of an irreducible whole participates in its wholeness."[23] The fundamental problem is that such a vision is incomprehensible to modern science and its reductionism. The notion of "being" has been jettisoned when it is actually indispensable to a fuller understanding of reality. Smith affirms that "being and wholeness are one and the same,"[24] and makes the important observation that "Different ways of knowing correspond to different kinds of being, or as we shall say, to different ontological domains."[25]

An obstacle in understanding wholeness, Smith explains, is that "We have become conditioned to think of wholeness in inherently set-theoretic terms"[26] which leads to reductionism. The notion of irreducible wholeness is proposed as an alternative to Cartesian bifurcationism. Among the spiritual traditions of the world, we find unanimous agreement that the tripartite constitution of the person—and of the cosmos of which we are but a mirror—comprises Spirit, soul, and body; or the spiritual, psychic, and corporeal states. Smith writes: "[M]an is tripartite—consisting of *corpus*, *anima*, and *spiritus*—so too is the cosmos at large;"[27] elsewhere, he writes: "To be precise: his body pertains to the corporeal realm, his soul or *psyche* to the intermediary, and his intellect or *nous* to the aeviternal."[28]

[22] "Physics and Irreducible Wholeness," *Physics: A Science in Quest of an Ontology*, 19.

[23] Ibid., 27.

[24] "Vertical Causation and Wholeness," *The Vertical Ascent*, 112.

[25] "*Sophia Perennis* and Modern Science," *The Wisdom of Ancient Cosmology*, 24.

[26] "Irreducible Wholeness and Dembski's Theorem," *Physics: A Science in Quest of an Ontology*, 45.

[27] "Science, Scientism, and Spirituality," *The Vertical Ascent*, 179.

[28] "Platonist Cosmology: The Long Lost Key," in *Physics: A Science in Quest of an Ontology*, 124.

The tripartite structure is a wholeness, not a division into unrelated parts. Smith observes that "the connection between 'spirit, soul, and body,' . . . is rooted in the *irreducible wholeness* of the *spiritus-anima-corpus* compound: a wholeness which actually derives from *spiritus* . . . the bond between *spiritus* and *anima-corpus* proves to be unbreakable."[29] The dominance of the unseen world proclaimed by the spiritual traditions (and their corresponding psychologies) was gradually overthrown by the more tangible demands of the sensorial world and its empirical modes of knowing, represented by the modern West's culture of materialism.

The cosmos consists not only of "being" (pertaining to all possibilities in the manifest order) but also of "non-being" (with respect to the unmanifest order). Metaphysics thus provides a synthesis between being and non-being, and our identity is fully integrated into the cosmos:

> [T]here is a symbolic correspondence between the cosmos and its entirety and man, the theomorphic creature who recapitulates the macrocosm within himself. Thus man is indeed a "microcosm," a universe in miniature, and that is the reason why, symbolically speaking, man is situated at the very center of the cosmos. In him all radii converge; or better said, from him they radiate outwards in every direction to the extremities of cosmic space. . . . No doubt the reason for this centrality is that man, having been made "in the image of God," carries within himself the center from which all things have sprung. And that too is why he can understand the world, and why in fact the cosmos is intelligible to the human intellect. He is able to know the universe because in a way it pre-exists in him.[30]

Smith's approach to the human psyche was grounded in a metaphysics that challenged the present-day reductionism which claims that the brain is everything; that is, "the psychic

[29] Ibid., 143.

[30] "'Progress' in Retrospect," in *Cosmos and Transcendence*, 142.

life is viewed as an epiphenomenon of brain function."[31] Furthermore, he writes that "it is the *soul* (in the traditional sense of *psyche* or *anima*) that transforms the material substrate into a living and sentient body. And this means that the living body is *more* than what the molecular biologist or neurophysiologist conceives it to be."[32] He adds, "a 'soul' belongs to a higher ontological order, and . . . consequently the ontological discrepancy between physical components and actual parts becomes correspondingly magnified."[33] It is because of the soul's intimate connection to the transpersonal order that we have sensation, perception, thoughts, and consciousness.

An understanding that grasps both the horizontal and vertical aspects of existence is necessary to fully grasp the nuances of the human psyche; yet mainstream psychology, for the most part, confines itself to a purely secular outlook. As the vertical dimension pertains to unity, Smith writes: "horizontal causation cannot give rise to irreducible wholeness."[34] He provides the following key to discerning the horizontal from the vertical: "Vertical causality derives from wholeness, whereas the horizontal derives from parts . . . ontologically speaking, the whole has primacy over the parts."[35]

From the post-medieval West to the present, there has been a conflation of the "eye of the heart" (the transpersonal Intellect) with the faculty of reason. It is through this noetic faculty that we can know the Real. Smith writes that "the essential precondition to all such "seeing" is doubtless a radical *metanoia*: a shift of our intellective gaze from the outward or sense-perceived world to the inner."[36] The predicament that afflicts modern psychology is precisely this: namely, seeing

[31] "Rediscovering the Corporeal World," *TQE*, 24.

[32] "The Enigma of Visual Perception," *Science and Myth*, 93.

[33] Ibid., 94.

[34] "Irreducible Wholeness and Dembski's Theorem," *Physics: A Science in Quest of an Ontology*, 56.

[35] "The Tripartite Wholeness," *The Vertical Ascent*, 18.

[36] "Neurons and Mind," *Science and Myth*, 129.

the human psyche on a purely profane level, cut off from its "vertical" dimension, which alone can resolve the debilitating impasse that this discipline has brought upon itself.

It needs to be said that "to the extent that the mind has been purified—the "mirror" wiped clean—the "seeing" takes care of itself." [37] As the Divine is both transcendent and immanent, when our integral vision is restored, we are able to perceive the vertical and horizontal dimensions respectively:

> Indeed, we could in no wise "sense" God outside of ourselves if he were not also present within the depths of our soul as the first and ultimate "seer." It is this inscrutable indwelling of God—as the "soul of our soul"—that enables and indeed powers the quest from its first inception to its ultimate end.[38]

It needs to be added that all sense perception is, ultimately, metaphysical in nature; it "is . . . a mystery—as are ultimately the entities we perceive."[39] Smith emphasizes the need to understand that "perception, like every other act of knowing, is consummated in a certain union or contact between the subject who knows and the object that is known."[40] Due to the dominant paradigm of reductive bifurcation, this truth has been obscured: "The harm derives mainly from an erroneous theory of perception, which had the effect of cutting the universe off from man, the human observer."[41] This has undermined our ability to know who we truly are, and our relationship to others and the cosmos: "We know the cosmos to the extent that we know ourselves; we are able . . . to know the outer world precisely because it corresponds to the inner."[42]

Smith presents a way of healing our infirm psyches, due to the spiritual crisis of the modern world:

37 "Metaphysics as 'Seeing,'" *Science and Myth*, 157.
38 Ibid.
39 "The Mystery of Visual Perception," *Vertical . . .*, 124.
40 "Cosmology in the Face of Gnosis," *Christian Gnosis*, 43.
41 Ibid., 44.
42 Ibid.

> The primary task of the true metaphysician is then to undo the collective decline, to reverse it in himself. It is a question of restoring the "heart" from its "darkened" condition, and in so doing, to recover the unimpaired use of our God-given "eyes": such, in brief, is the task of veritable metaphysics.[43]

To benefit from this awareness requires that we abandon the distortions of Cartesian dualism which "has had a profound effect on the Western psyche, to the point of provoking a kind of 'collective schizophrenia.'"[44] Our spiritual well-being, then, depends on moving beyond this dehumanizing aberration: "The defeat of scientistic materialism—the restoration of our collective sanity!—appears to demand nothing less."[45] Smith urges that "it is time to rediscover the integral cosmos."[46] Yet what can be done about this? He adds: "The remedy... is simple in the extreme: rectify this misconception—expunge the fallacy of bifurcation, root it out."[47]

It is imperative that we continue to dismantle, once and for all, this pernicious pseudo-ontology, as its spell has yet to be broken. Because "everything depends on how we perceive the world,"[48] we need to restore the "eye of the heart" to its rightful place, in order to rehabilitate the spiritual dimension to our understanding of mental health. In this way, we may be able to recover an authentic way of knowing and healing that is critical to humanity in these bewildering times.

Samuel Bendeck Sotillos

[43] "Metaphysics as 'Seeing,'" in *Science and Myth*, 158.

[44] Ibid., 160.

[45] "Three Vertical Powers of the Soul," *Physics and Vertical*, 32.

[46] "Cosmic versus Measurable Time," in *Physics: A Science in Quest of an Ontology*, 90.

[47] "Lost in Math: The Particle Physics Quandary," *The Vertical Ascent*, 57.

[48] "'Progress' in Retrospect," in *Cosmos and Transcendence*, 158.

Part Three

Physics and Metaphysics

Cosmology

Chapter X

The Restored Foundations

I did not know Dr. Wolfgang Smith very well, but I consider myself privileged to have had several exchanges with him over a period of several years. What impressed me the most about him was his purity of heart and his simple love for the truth, against the backdrop of an academic universe in which human respect and the desire for fame and fortune often exert a paralyzing effect on the pursuit of Truth.

In 2019, Dr. Smith participated in the production of a video series produced by the Kolbe Center for the Study of Creation entitled "Foundations Restored" in which he demonstrated that recent discoveries in astronomy had falsified the Copernican Principle, so dear to Enlightenment thinkers, as well as the Big Bang hypothesis. In Episode 8 of the DVD series he stated:

> In recent years, astrophysical observations have detected—and subsequently verified beyond all possibility of doubt—a global cosmic structure which disproves the Big Bang theory definitively, that is to say, once and for all. Whether the leaders of the astrophysics community are prepared to admit the fact or not, the theory is now defunct. For all its vaunted accomplishments and Nobel Prize discoveries—which now prove not to be real—it too has turned out at last to be no more than a godless ideology masquerading in scientific garb.
>
> What, then, are these findings, and why do they prove fatal to the theory?
>
> That is what I wish now to explain.

It is to be noted, first of all, that Einsteinian physics—that is to say, the Einsteinian deviation from the classical theory—is premised on the stipulation that the cosmos is bereft of design, namely, of structures that cannot be explained in terms of random fluctuations. Unacknowledged though it be, it is upon this ideological assumption for which there is no scientific support at all that Einstein's so-called Principle of Relativity and thus Einsteinian physics at large is actually based. Now, in the case of Big Bang cosmology that stipulation takes the form of the so called Copernican—also termed Cosmological—Principle, which affirms that, when viewed on a sufficiently large scale, the cosmos proves to be perfectly uniform.

It is supposedly with the cosmos at large as with a volume of gas. Whereas, on a microscopic scale, there are molecules moving at different speeds and in different directions, yet when viewed on a large scale the gas is nonetheless homogeneous and fully specified by its temperature, pressure, and density. Inasmuch, therefore, as Big Bang cosmology satisfies the aforesaid Copernican Principle, the discovery of even a single large scale structure—a structure, that is, which does not conform to that Principle—suffices to disprove the theory. And this is precisely what has happened within the last decade.

Meticulous and incontrovertible observations, carried out by a satellite—named after the physicist Max Planck—have revealed a global structure, extending through the cosmos at large, a fact which flatly contradicts the Copernican Principle.

But actually there is more—a great deal more!

Let me fill in the specifics.

What has been discovered is that there is an electromagnetic radiation, identifiable by its spectrum, filling the cosmos at large, known as the cosmic microwave background or CMB. Aficionados of Einsteinian cosmology regard it as a remnant of the universe some Three Hundred Thousand

> years after the Big Bang. In truth, however, it is this structure in the CMB which rigorously disproves that Einsteinian theory by the fact that it constitutes an axis—a great circle in fact—something which, most assuredly, should not be there, something namely that cannot be explained away as the result of statistical fluctuations. As we have said, the existence of that axis disproves Big Bang cosmology.
>
> Period.
>
> Yet there's more!
>
> The most amazing fact of all utterly devastating to the Einsteinians is that this axis constitutes a circle within the ecliptic of our solar system! That solar system, which was supposed to resemble a random flake of dust in relation to the cosmos as a whole, proves thus to be special. So highly special, in fact, as to determine the global structure of the universe! Nothing could be more antithetical not only to Einsteinian physics, but indeed to post-Galilean astronomy at large. This discovery of what astrophysicists have, quite understandably, dubbed "the axis of evil" actually points us back to the brink of pre-modern astronomy, and signals thereby the end of what historians term the Enlightenment.

The most wonderful part of this presentation was the childlike way in which Dr. Wolfgang revealed the nakedness of the Emperor of Enlightenment Philosophy, whose New Clothes woven of naturalistic explanations had been exposed as transparent fallacies, no longer capable of concealing the Creator's magnificent design for man and the universe.

Hugh Owen

Chapter XI

Pondering Wolfgang Smith's Platonist Cosmology

In its contemporary usage, the term *cosmology* refers to a highly specialized branch of physics that is dedicated to understanding the universe at its grandest of scales. At the heart of twenty-first century cosmology lies the *Big Bang Theory*. Today's official creation story, which states that the universe began with a *singularity*; the illusive point where the Einstein field equations, which describe space-time, "break down" beyond any mend.

Although modern cosmology is indeed more august mathematically, and its practitioners lauded as the savants of today, the field nevertheless represents only a small fraction of a much wiser intellectual tradition. Long before talks of CMB maps, redshift, inflation, and dark matter there stood, of course the ancient philosophers, whose respective cosmologies were not limited to what Lord Kelvin called "the science of measurement"—i.e., to physics. But, instead, sought to grasp the universe, in its wholeness, using metaphysical and even spiritual modes of enquiry.

Among the earliest of thinkers from the ancient tradition was Thales of Miletus. Often considered "the first philosopher,"[1] Thales famously proposed that the Earth rests on water and that water itself is the fundamental principle or *arche* of all things.[2] While today such ideas might appear absurd, Thales nonetheless marks the inception of the Western philosophical

[1] O'Grady, "Thales of Miletus," IEP (Internet Encyclopedia), Introduction, https://iep.utm.edu/thales/.

[2] Ibid., Section 3.

tradition. A tradition that arguably reached its apex in Plato's cosmology and, significantly for this book, serves as the foundation that underlies the cosmology of Wolfgang Smith.

Accordingly, in this chapter I will attempt to explain Smith's cosmology in depth, while also interweaving my own insights on its general significance. To this end, I will start by outlining Plato's original cosmology, the foundation upon which Smith builds his. Following this I will explain and explore the three domains that comprise Smith's cosmos. At this chapter's conclusion, I will wrap up by examining how the three domains of Smith's cosmology hang together through participating in *Irreducible Wholeness* (IW).

Plato's Cosmology

At the core of Plato's original cosmology, found in his *Timaeus*, is the notion of an unchanging realm (*eidos*) outside of space and time.[3] This domain, which the eighteenth century Neo-Platonist Thomas Taylor called the *Intelligible Realm*,[4] is populated with forms.[5] In short, forms are intelligible blueprints or ideals which material objects strive, fallibly, to imitate. A golden sphere, for example, attempts to replicate the ideal of sphericality. However, in striving to manifest this ideal, the sphere inevitably gets incapacitated by the limits of its own matter. And, thus, perpetually falls short of attaining pure roundedness.

An additional but equally central facet of Plato's cosmology is the world-soul (*anima mundi*).[6] In short, this cosmic soul functions as an arbiter between the world of Forms and the ever-changing sensory world[7]—the realm of green grass, blue skies, and all other perceptible phenomena. In essence, Plato's

[3] *Timaeus*, 50d–52b.

[4] *The Works of Plato*, Vol II, Thomas Taylor (London: R. Wilks, Chancery-Lane, 1804), 417.

[5] *Timaeus*, 50d–52b.

[6] Ibid., 30c–34a.

[7] Ibid., 30c–34a.

world-soul is a non-divine superintelligence that emanates and animates the visible universe, shaping it to mirror the perfection of the Forms up above.[8] Intriguingly, St Augustine of Hippo, along with several other early Church Fathers,[9] appears to have embraced Plato's idea of the world-soul.[10]

Yet still, St Augustine went one step further than Plato. For he asserted that beyond the world of Forms and the sensory world there reigns the one true Christian God, the divine originator of the Forms themselves.[11] In contending so, St Augustine effectively Christianized Plato's cosmology, a transformative move that had a long-lasting impact on generational thinkers to come, including Wolfgang Smith.

Admittedly while Smith's cosmology does differ somewhat in its outward expression from Augustine's and Plato's, at its esoteric core Smith's cosmos is exactly isomorphic to the Platonic cosmology.[12] Since, in essence, it articulates a vision that aligns with the deeper, more mystical interpretations of Plato's teachings—those said to have been revealed only to an inner elect within the *Academy*.[13] This particular reading of Plato, rooted in more esoteric teachings, is known as Neoplatonism,[14] a school of thought founded by Plotinus,[15] the great third century Esotericist, and one followed closely by Smith.[16]

[8] Ibid., 30c–34a.

[9] Travis Dumsday, "Breathing New Life Into the World-Soul? Revisiting an Old Doctrine Through the Lens of Current Debates on Special Divine Action." *Modern Theology* 35 (2019): 301–22: https://onlinelibrary.wiley.com/doi/abs/10.1111/moth.12448.

[10] Teske, Roland J., "The World-Soul and Time in St. Augustine," *Augustinian Studies* (1983) 14:75–92.

[11] St Augustine, *The Confessions.*

[12] Smith, PSQ, Part III, chap. 1.

[13] Ibid.

[14] Christian Wildberg, "Neoplatonism," SEP (The Stanford Encyclopedia of Philosophy) (Winter 2021 Edition), Edward N. Zalta (ed.): https://plato.stanford.edu/archives/win2021/entries/neoplatonism/.

[15] Ibid., Section 1.

[16] Smith, PSQ, Part III, chap. 1

While the intricacies of Neo-Platonism are far too elaborate to explore in this essay, it is enough to note that its cosmology is structured around three main ontological levels. These levels, as described by Smith are the Aeviternal, the Intermediary, and the Corporeal.[17] As a brief aside, although the concept of the "world-soul" does figure heavily in traditional Neo-Platonic thought, it is not explicitly included in Smith's formulation. In a previous chapter, I discussed how the world-soul might be integrated into Smith's metaphysical framework. However, rather than lean further into that speculation, I will turn instead to analyze how these three ontological tiers converge to form the cohesive and organic cosmology that underlies Smith's vision of reality. With that purpose, in mind, I will now proceed to examine the domains of aeviternity, the intermediary, and the corporeal.

The Aeviternal Realm

At the foundation of Smith's cosmology lies the aeviternal plane—Smith's first ontological level—which very broadly corresponds to Plato's intelligible realm.

At its core, the aeviternal plane aligns with a concept introduced by Smith in his later writings: namely Irreducible Wholeness.[18] As Smith so often noted, IW signifies a wholeness that transcends a sum of parts.[19] In contrast to ordinary wholeness, which is simply a sum of parts, crucially, Wolfgang Smith regards IW as existing independently, unbound to any particular manifestation of something physical.[20] In fact, Smith sees IW as the source of all Being in the cosmos.[21] A pivotal idea we shall revisit when exploring IW's implications.

[17] Smith, PSQ, Part III, chap. 1, 99.
[18] Ibid., 23.
[19] Ibid., 45.
[20] Ibid., 23.
[21] Ibid., 28–29.

Furthermore, Smith's aeviternity also contains various *modes of being*. Modes that derive directly from IW.[22] In his 2004 book, *Ancient Wisdom and Modern Misconceptions*, Smith introduces a notable mode of being, pertaining to aeviternity, which he termed *celestial corporeality*.[23] This modality of aeviternity refers to the bodily nature and extension of a thing in its angelic or higher form—essentially corresponding to what we perceive as visible light.[24] While a full exposition of celestial corporeality would vastly exceed the scope of this essay, it is worth noting that visible light remains a profound mystery unto itself. A mystery that, if adequately explored, might, plausibly, yield further insights into Smith's notion of vertical causation.

As stated by Arthur M. Young,[25] a man whose life shared many parallels to Smith's, light functions as a sort of "mediator that collapses possibility into actuality."[26] In turn, Young's compelling claim, here, raises some intriguing questions in view of Smith's metaphysics: Could celestial light be a kind of custodian of vertical causation? And if so, what is the relationship between physical light and celestial light that actualizes?

These are thought-provoking questions, indeed, ones that arguably deserve deeper exploration—perhaps in a dedicated essay. For now, however, I leave them as open questions to be explored!

Getting back on track, a far more familiar modality of *Aeviternity* is the elusive present moment.[27] In agreement with St Thomas Aquinas, Smith held that the present moment exists in some sense *above* time as a *nunc stans* (*standing now*).[28] It exists, that is, as a still and unmoving pivot point around

22 Smith, *Ancient Wisdom and Modern Misconceptions*, 68.

23 Ibid.

24 Ibid., 81.

25 They both worked at the Bell Aircraft Corporation in the 1950s and were both fascinated with Whitehead from an early age.

26 See Young, *The Reflexive Universe*.

27 Smith, "The Mystery of Visual Perception."

28 Ibid.

which the flow of temporal succession unfolds. As such, the present moment can be understood as a modality of the aeviternal realm, participating in its timeless character while also remaining perceptible within temporal experience.

However, for the attentive reader, several questions naturally arise at this juncture. For one, how can we arrive at such conclusions concerning the nature of the present moment? Why should we conclude that the present moment is outside of time specifically? Secondly if the present moment is indeed atemporal, why does it also seem to be so deeply embedded within time?

In answer to the first questions, demonstrating that the present moment is outside of time proves to be surprisingly straightforward. A proof of this *ontological theorem* follows naturally from the fact that we can simultaneously associate the present moment with different durations of time. For instance, as I am writing this, I can simultaneously identify myself as being present, i.e., right now living in: the 27th day of June, the year 2025, and the decade of the 2020s. And just like that the present moment can simultaneously be associated with multiple distinct durations of time, each vastly different in length.

However, if the present moment can simultaneously be identified with multiple distinct durations of time it follows that the present cannot subsist within any one of these durations absolutely. That is, the present moment is not to be found within any one length of time over another, no matter how small or gargantuan that length is. For, if it could be this would contradict the fact that we can, within that same period, also identify the present moment with another temporal duration—be it smaller or larger. On a Wednesday morning, for example, I can identify the present moment with that day, but I can also extend it to the specific week or month I am going through. On the opposite end of the spectrum, I can also identify the present with a specific minute or even millisecond, on that same Wednesday morning. Ergo, the present moment cannot be pinpointed to one length of time absolutely, which, in truth, means that the present must reside outside of time altogether because it exhausts all intervals of time in its identification. As the Angelic Doctor says *ad nauseum*, "time consists of befores and afters";[29] the present, therefore, must be above time!

In response to the second question—why the present moment seems so deeply submerged in time if it also exists above time—I would speculate this is because while the present is indeed above time it can also descend, if you will, into temporality. And, in fact, has to given Wolfgang Smith's understanding of time, which will be explicated shortly. As stated previously, the present moment pertains to IW. This means that presentness can descend into time without losing its transcendent origin. Just as the ideal of circularity can descend into the material world, by participation, without fully leaving aeviternity, so too can the present descend into time without departing from its non-temporal origin. In other words, the total irreducibility of the present to any of its specific instantiations gives it the kudos to be in two places at once: in aeviternity in its absolute form and in time in its relative or participatory form.

Moving onward a third, and final, modality of aeviternity, in Smith's cosmology, is what Plato terms *the realm of forms*. Plato's forms we might say are an expression of IW and encompass all truths graspable with the intellect.[30] To be grasped or understood by the intellect, instead of rationality, is for a truth to be recognized by the soul, instantaneously, without any sequence of reasoning, *from premise to premise,* so to speak! I can recognize, for instance, that nothing can both be and not be, at the same time, without any justification.

The forms, graspable with the intellect, range vastly from the basic truths of classical logic, such as the principle of non-contradiction, to the basic truths of mathematics and

[29] Aquinas, *Summa Theologiae*, Prima Pars, Question 10, Article 5.
[30] Smith, PSQ, Part III, chap. 1.

geometry.[31] These logical and mathematical forms exist as expressions of aeviternity beyond our sensorial perception.

Anyhow, the key point to take away from this section, before moving on, is that the aeviternal realm corresponds to IW at its most fundamental core. Irreducible wholeness, and resultingly aeviternity, possess various modalities. These modalities include: Celestial Corporeality, The Present Moment and intelligible forms to name a few!

The Intermediary Realm

The second level of Smith's cosmology is: *the intermediary realm*. Drawing on Aquinas again, Smith aligns himself with the view that time exists independently of space.[32] An axiom which canonically defines the intermediary realm as the domain where time is present, but space is absent.[33]

For Smith, time possesses dual ontological modalities. The first mode is measurable time—a sequential, quantifiable progression (t_1, t_2, t_3, etc.). The second is cosmic time.[34] Cosmic time, Smith argues, is an "active principle of cosmogenesis,"[35] a generative force that impresses the limits of beginning and end upon the otherwise aeviternal present. In this view, the present moment descends into temporality, acquiring both a before and after from cosmic time.[36] Resultingly, this ontological descent gives rise to events, i.e., to bounded segments of time that structure our experience of becoming.[37]

Hence, in a sense *time is the scattering of eternity*. Insofar as time is a process whereby the present moment, a mode of aeviternity, totally beyond time, continuously receives a

[31] Ibid.
[32] Ibid., 99.
[33] Ibid.
[34] Smith, "Cosmic versus Measurable Time."
[35] Ibid.
[36] Ibid.
[37] Ibid.

new before and after. Initially, on a global scale cosmic time determines the longest event, which we are currently going through. Namely, the before and after of the entire temporal order—the beginning and end of the cosmos. This entire cosmic event is determined from the outset by cosmic time, acting on the atemporal present. However, the events that happen, in between, are not. Thus, cosmic time is continuously dispensing a set of miniature befores and afters to "fill in the gaps." As tripartite beings, who pertain to aeviternity but also to the lower realms of the cosmos, we are inevitably propelled along this stream of miniature durations; with each moment we encounter a refracted glint of eternity. Time, then, is not simply the ticking of clocks but the medium through which eternal realities shimmer into temporal form. We move, not just forward, but inward through veils of becoming that both conceal and reveal the timeless within the temporal. In this spirit, time is the scattering of eternity!

By defining time, as the *scattering of eternity* Smith's cosmology, lends credence to the common phenomenal experience of time, wherein the present moment stands as still as a point, but also as a fundamental unit of time itself—a unit which I have argued is being continuously scattered.

Additionally, the bounds of cosmic time, which initiate this scattering, are not actually measurable in themselves.[38] But rather are bounds that give rise to events that can be measured.[39] Importantly, the actual measurement of duration comes from the operations of physics or quantification. To be more precise, cosmic time sets a specific duration, however, this duration is qualitative and immeasurable[40] and is, perhaps, equivalent to what Henri Bergson called *la durée*—that is pure duration unaffected by measurement.[41] This pure

[38] Ibid.
[39] Ibid.
[40] Ibid.
[41] See Bergson, *Time and Free Will.*

duration, which can only be fully comprehended in conscious experience, can also be defined measurably under the units that clocks measure with.[42]

Beside from being the domain of time, the intermediary also acts as a sort of "psychical realm." A realm wherein we can use our imagination or *phantasia* to construct objects that take us to the intelligible truths of aeviternity.[43] In the penultimate chapter of his final book, based on the writings of Thomas Taylor, Smith argues that the geometer, for example, constructs imaginary objects that arouse *the energies of his intellect.*[44] With these energies the geometer can reach conclusions, he wouldn't have reached otherwise about geometric shapes.[45] For example, the geometer can imagine a circle in his mind when seeking out its associated implications. Crucially, the imagined circle of the intermediary is not the same as the circle of aeviternity, for the former has imperfections by dint of simply being imagined whereas the latter does not.[46] Nevertheless, the imagined circle is the closest thing to aeviternal circularity that we can encounter visually. Aeviternal circularity, on the other hand, is the form or essence of circularity totally divorced from any instantiation.[47]

The Corporeal Realm

The final realm of Smith's tripartite cosmos is the corporeal world—the domain of sensory qualities in which we live, move, and have our being.[48] By "sensory qualities," Smith means attributes perceivable with the five senses. These attributes include colors, sounds, tastes, textures, and so on. Since the

[42] Ibid.
[43] Smith, PSQ, Part III, chap. 1, 106.
[44] Ibid., 105.
[45] Ibid., 101–2.
[46] Ibid.
[47] Ibid, 107.
[48] Smith, *TQE*, chap. 1.

early modern era, philosophers have largely dismissed these qualities as fantasies of the mind, not truly "out there" in the external world.[49] Smith, however, firmly rejects this spurious premise. He contends that sensory qualities are not just mental paint but genuine parts of external "corporeal objects" e.g., a red apple.[50]

Complementing the corporeal world is the physical universe, the realm investigated by the physicist. This realm includes subcorporeal objects (SX), which are corporeal objects (X) understood in purely physical terms,[51] and transcorporeal objects, which are physical entities not associated with any X.[52]

Smith's primary motivation for reintroducing the corporeal realm is its ability to address the long-standing quantum measurement problem.[53] To this end, Smith proposes that corporeal reality introduces a non-physical yet ontologically real component to the objects of classical physics, namely their sensory qualities.[54] This addition is not merely aesthetic but instead plays a vital role in halting the so-called Von Neumann chain.[55]

In standard quantum mechanics, measurement leads to an infinite regress in which: the measuring device becomes entangled with the system, the observer with the device, and so on, with no clear end point at which the system "collapses" into a definite state.[56] Smith argues that the corporeal realm, the realm of concrete sensory perception, breaks this chain by introducing a level of qualitative definiteness that is not present in purely physical systems.[57]

49 Ibid., 7.
50 Ibid., 8.
51 Ibid, chap. 2, 34.
52 Ibid., 35.
53 Smith, *PSQ*, chap. 2, 14–15.
54 Ibid.
55 Ibid.
56 *SEP*, "Philosophical Issues in Quantum Theory," chap. 4.
57 Smith, *PSQ*, chap. 2, 14–15.

Therefore, when a quantum system is confronted with the definiteness of objects containing sensory qualities, for example a pointer reader, this necessitates a transition from quantum indeterminacy to classical determinacy.[58] In other words, it engenders wave function collapse and terminates the so-called Von Neumann chain.[59]

Beyond offering a resolution to the measurement problem, I am personally convinced that Smith's account presents a fresh and compelling way to reimagine the problem in the first place. Traditionally, the measurement problem in quantum mechanics has been framed as the puzzle of how quantum systems, described by probabilities, "collapse" into definite outcomes when measured.[60] Crucially, this issue has always been approached from a physical standpoint. In response, various physical mechanisms have been proposed to account for the transition from the indeterminate realm of quantum probabilities to the concrete world of observed values.[61]

What Wolfgang's solution highlights, however, is that there are actually two distinct problems at play. Borrowing terminology from David Chalmers' reflections on consciousness, we can think of these as the easy and hard problems of quantum measurement.

The easy problem concerns the quest for a physical mechanism that explains how the wave function collapses, or appears to collapse, into a definite state. To this end, physical solutions have been proposed ranging from GRW collapse[62] to more recent labors by Drossel and Ellis to explain collapse using the physics of classical heat baths.[63] The hard problem, by contrast, asks the question of whether there ought to be a deeper

[58] Ibid.

[59] Ibid.

[60] *SEP*, "Philosophical Issues in Quantum Theory," chap. 4.

[61] Ibid.

[62] See *SEP*, "Collapse Theories."

[63] Drossel and Ellis, 2018, "Contextual Wavefunction Collapse."

non-physical reason as to why the state vector collapses to begin with? The answer to the hard problem depends very much on one's ontology. If we consider quantum systems, for instance, to be *potentiae* then no amount of physical explanation can answer their transition from one ontological state to another. This is why the corporeal world is absolutely mandated for Smith's solution to work. If we don't, however, then the jury is still out!

Apart from its connection to the quantum measurement problem the corporeal domain Smith argues is supported by the work of James Gibson, a perceptual psychologist who worked for the United States Army Air Corps during WWII.[64] According to Gibson, who supported his work with experimental rigor, visual perception is mediated through the ambient optic array, the structured pattern of light that exists at a point of observation.[65] Within this array are invariants, stable properties of ambient light that, according to Gibson, convey sensory information to the observer.[66] From Gibson's perspective, sensory qualities are not internal constructs but are externally existent features of the perceptual environment. This view aligns with and reinforces Smith's notion of a corporeal world with real perceivable properties.[67]

What fascinates me about Gibson's theory, especially, is that it aligns perfectly with Aquinas's theory of visual perception. According to Aquinas sensory qualities are carried as forms through atmospheric light. The forms trapped in light Aquinas calls *species in medio*.[68] Evidently, Aquinas's

[64] Smith, *PSQ*, 10.

[65] Gibson, 1979, chap. 5.

[66] Ibid.

[67] Ibid.

[68] Eleonore Stump, "Aquinas on the Mechanisms of Cognition: Sense and Phantasia." In *Medieval Analyses in Language and Cognition*, edited by Sten Ebbesen and Russell Friedman, 377–95. In the series *Det Kongelige Danske Videnskabernes Selskab, Historisk-Filologiske Meddelelser 77* (Copenhagen: Munksgaard, 1999).

theory comports with Gibson's since both assume that light contains sensory qualities.

As we come to better understand the corporeal world a synthesis of Aquinas and Gibson could be very helpful in clarifying the distinction between sensory qualities as they pertain to external objects, and sensory qualities as they pertain to light.

From Irreducible Wholeness to God

Having outlined the three realms of Smith's tripartite cosmos, we can now turn to the question of how these domains cohere into a cosmology. At the beginning of this essay, I implied that a cosmology in the ancient sense is a metaphysical and even spiritual system of the universe, as a whole.

Central to Smith's particular cosmology is Irreducible Wholeness, the metaphysical glue that binds the cosmos together. As we have previously seen, IW constitutes the most fundamental mode of aeviternity from which the rest of the cosmos follows. Hence, everything that exists within the cosmos must, in some way, participate in or refer back to this principle. In this sense, the cosmos itself can be understood as an expression or embodiment of IW. An assertion that, incidentally, is supported by Thomas Taylor's exegesis of Plato, in which he describes the cosmos as a "whole mundane animal."[69]

Since the cosmos is an IW, it follows that each of Smith's three domains contains traces or reflections of the others. That is, within any one realm, the presence of the others can be discerned, if we dare to look closely enough. For example, in the aeviternal realm we can find traces of the corporeal in celestial corporeality. And in the intermediary domain we can find traces of aeviternity in the descending present moment.

However, the fact that each domain is itself an IW and that, collectively, they constitute a distinct overarching IW, poses a significant ontological knot to be untied. Generally

[69] Thomas Taylor, *The Works of Plato*, 415.

speaking, I believe it is correct to say that IW as a concept precedes IW in its instances. That is the IWs found within the modalities of aeviternity, the intermediary, and the macrostructure of the corporeal world are ancillary to IW *simpliciter* which lives, if you will, in the aeviternal plane unattached to its instantiations. However, precisely how these IWs connect is something that needs to be thought out far more deeply beyond Smith's current analysis.

I began this chapter by contemplating the state of modern cosmology—a field undeniably obsessed with quantification. With the cosmology of Wolfgang Smith, in hand, we now possess the power to challenge this *Weltanschauung* quite forcefully. And to hopefully restore an understanding of the cosmos that goes far beyond relations between spatio-temporal parts.

Ultimately, such an understanding can guide us back to God, the creator and sustainer of the cosmos we inhabit.

John Taylor

Part Four
Science, Metaphysics, and Religion

Chapter XII

Response to Wolfgang Smith[1]

Professor Wolfgang Smith is one of the very few scientists who have devoted their lives to the pursuit of science and are at the same time rooted in the teachings of the perennial philosophy. Therefore his essay "Sophia Perennis and Modern Science" is very pertinent and reveals the essential truth that the tenets of perennial philosophy are not only significant in the domains of religious studies, traditional art, psychology, and the like. Perennial philosophy is also of the greatest importance for a revaluation of the philosophy of the modern sciences and for providing a meaningful framework for the understanding of these sciences and especially what constitutes their basis, namely, quantum mechanics. This latter point has been already treated in Smith's remarkable opus, *The Quantum Enigma*, and in fact much of his discussion in this essay is related to the theses of that work, although the present essay is an exceptional synthesis of Smith's thought on the subject of the relation between the perennial philosophy or sophia perennis and modern science and not simply a summary of *The Quantum Enigma.*

I am in such deep agreement with nearly everything written in this essay that there is hardly a point which I would wish to criticize. My response on most issues will be in fact simple confirmations. Nevertheless, there are a number of points upon which I wish to expand in order to clarify further my own views on the subject. Smith writes that "science in the

[1] This contribution first appeared as "Reply to Wolfgang Smith" in *The Philosophy of Seyyed Hossein Nasr*. Copyright © 2001 by The Library of Living Philosophers. Reprinted by permission.

traditional sense is thus a matter of 'reading the icon'—a far cry indeed from the Baconian vision!" This is a very apt manner of speaking of the traditional cosmological sciences. These sciences depict a cosmos which revealed a meaning beyond itself and can be contemplated as an icon. One could in fact go a step further and say that, using the language of Christianity, the cosmos is an icon and can only be understood in depth as an icon, which reveals a divine reality beyond itself. The traditional cosmological sciences brought out this iconic reality and permitted those who studied and understood them to see the cosmos as an icon and to be able to contemplate it rather than knowing it only discursively. The modern sciences, issuing from what Smith calls "the Baconian vision," also know nature but no longer as an icon. They are able to tell us about the size, weight, and shape of the icon and even the composition of the various colors of paint used in painting it, but they can tell us nothing of its meaning in reference to a reality beyond itself. What they tell us about the size, composition of the paints, and so on, of the icon are not false on their own level, but they do not exhaust knowledge of the icon and it would be both ignorance and hubris to claim that this type of knowledge is the only knowledge possible of the icon. The consequences of this ignorance parading as totalitarian science combined with hubris, made even more dangerous by being denied, are lethal for man's spiritual life. This ignorance, hubris, and denial have dire consequences even more outwardly in man's relation to the world of nature, divorced in the modern world from ultimate meaning as a result of the exclusive claims of Baconian science.

Smith qualifies his comments about Baconian science by saying that contemporary science at its best is not completely Baconian as Einstein's occasional comment about the "Old One" suggesting that he too may have been searching for "vestigia of a kind" shows. I agree that there are a number of individual scientists, even in the contemporary period, who, like the English botanist John Ray, cultivated science with the goal of discovering the *vestigia Dei* in creation. But they were and are functioning within a scientific framework in which such concerns by an individual scientist could not in any way affect the science they have produced. One can study and accept the theory of relativity with or without references to the "Old One," which means that the result of Einsteinian science is not the search for the *vestigia Dei* and the depiction of the cosmos as an icon, whatever may have been Einstein's personal views and attitudes.

The distinction that Smith makes between what ought to transpire in the mind of a modern scientist, which is reasoning upon data provided by the senses, and the meaning of knowledge according to the sophia perennis is of crucial importance. The epistemology provided by the sophia perennis covers "an incomparably greater range of cognitive possibilities" to quote the author, since it relates all acts of knowing to participation of the human intellect in the light of the Divine Intellect. In much of the discussion going on today in the domain of epistemology, both scientific and otherwise, this issue is forgotten or at least not emphasized, including the works of many contemporary Muslim thinkers writing on the subject. This participation is not confined to "moments of illumination" but involves all knowledge which relates the human subject to the object that is known. Smith quotes the Gospel of John and states, "what ultimately connects the human subject to its object in the act of knowing is indeed 'the true Light which lighteth every man that cometh into the world'"(1:9). Since the sophia perennis is both perennial and universal, one needs to add here how central this thesis is in other traditions. Being better acquainted with the Islamic tradition than with others, I can tum to that tradition and add that there are numerous verses of the Quran and ḥadīth about the relation of knowledge (*al-'ilm*) to light which constitutes a vast hierarchy issuing from God. In fact, according to the Quran (XXIV:35) God is the light of not only the heavens but also the earth. On the basis of these traditional sources and certain elements of Greek philosophy, Islamic philosophers, going back to al-Farabi and Ibn Sina, spoke of the illumination of the human intellect by the Active Intellect in the act of intellection. The symbolism of light was particularly central to the teachings of Suhrawardi, the founder of the School of Illumination (*al-ishraq*) and what Professor Smith has written on participation in the Light of the Divine Intellect is practically identical with the views of Suhrawardi, except that the latter identifies this light in its various degrees with the different angelic substances. Again I need to emphasize how important this issue is for an in-depth study of the epistemology of modern science in light of traditional teachings, and as far as Islamic thinkers now writing on the subject of epistemology are concerned, how crucial it is first to understand the tenets of traditional Islamic philosophy in this domain before embarking upon often puerile and moreover fruitless comparisons of traditional epistemologies and modern scientific epistemology.

Smith points quite correctly to the need for a living sapiential tradition within which cosmological sciences are cultivated and without which they wither and die "thus giv[ing] rise to what may indeed be termed a superstition." This is a very correct assessment upon which I must elaborate. A superstition is literally something whose ground has been removed. The metaphysical teachings of the sophia perennis constitute precisely the ground upon which the traditional cosmological sciences stood. Therefore, with the destruction of that ground these sciences could not but be reduced to superstition, although they still carried residues of truths no longer understood. The whole phenomenon of occultism in the West is quite interesting from this point of view. In other civilizations where the metaphysical ground has not been destroyed, there are certainly forms of popular superstition which are also very much present in the modern world, albeit in other guises, but in those civilizations one does not encounter the phenomenon of occultism as it developed in the salons of France and elsewhere in Europe from the eighteenth century onward. Many modern people, especially those of a scientific bent, immediately dismiss the traditional sciences such as alchemy as being simply superstition. But they do not realize that such sciences are like jewels which glow in the presence of the light of a living sapiential tradition and become opaque once that light disappears. Paradoxically enough, by claiming to relegate the traditional cosmological sciences simply to the category of superstition, the modern scientific enterprise has not only been helpless before the mushrooming of interest in these traditional sciences even in their residual form known in occultist circles, but it has been instrumental in the rise of new forms of superstition, such as the idea of progress, which are much more dangerous for the future of humanity than the practice of predictive astrology.

Returning to the question of perception, Smith emphasizes that light waves, sound waves, brain function, and the like are of course necessary and play a role in the act of perception but that "they do not—they cannot! constitute the perceptual act" which is "literally not of this world." He also lays the error of belief in bifurcation and the lure of it at the feet of the state of forgetfulness of "participation" by post-medieval European man. I have restated these lines to emphasize their central importance and my full agreement with it. Modem philosophy, psychology, or science are simply not able to explain perception which they always reduce to one of its parts or something else because the participation of the human intellect in the Light of the Divine Intellect is simply beyond the truncated worldview within which all modern thought, whether it be philosophical, psychological, or scientific operates. The rediscovery of the real significance of perception is only possible in light of the sophia perennis and is itself a key for the discovery of the metaphysical universe depicted by the perennial philosophy in its vastness and wholeness. And I agree with Smith that the greatest obstacle to the integration of modern science into the higher orders of knowledge and the rediscovery of how the miracle of perception works is Cartesian dualism or the theory of bifurcation.

I believe that the ingenious distinction made by Smith between the corporeal and physical worlds and the confining of quantum mechanics to the physical rather than the corporeal world, as well as the relation between the two of which he speaks, are major steps in the formulation of a more meaningful philosophy of physics in accordance with the sophia perennis. Like Smith, I also wish to emphasize that the corporeal and the physical worlds, as defined by him, are not only different but constitute two ontologically distinct domains. There is an ontological hiatus between the two and one cannot say that this physical stratum contains simply the "building blocks" for the corporeal world. The whole idea of fundamental particles from which we can build up the corporeal world with its forms and qualities is therefore false; and form and quality associated with the corporeal world—not to speak of psychological and spiritual realities—can never be reduced to the quantitative elements of the physical world which alone can be studied through the "modern scientific method" and be made to constitute the subject of quantum mechanics. Naturally, without mathematics there is no possibility of study of that physical world, as defined by Smith. To understand the ontological status of the corporeal and physical worlds would also solve the status of quantum physics not only in its relation to the world of classical physics but also to the corporeal, or what is ordinarily called the "physical world" (contrary to Smith's terminology), and the traditional sciences which deal with the qualitative and formal aspects of that world. This ontological awareness would also make clear the basic point I have mentioned in so many of my writings, namely, that the traditional cosmological sciences are not simply crude attempts to understand nature and primitive stages of the modern quantitative sciences, but contain profound knowledge of the formal and qualitative aspects of the corporeal world not reducible to quantity nor of lower significance than quantity *à la* Galileo. On the contrary they refer to realities with a higher ontological status than the quantitative.

The analogy made by Smith between the *materia secunda* and the Euclidean plane from which geometric forms emerge and in which they become actualized is also a brilliant one with which I could not but agree. It is so important to remember that the objects of modern physics are not like tables and chairs except smaller, but that they are "constructed" to quote Smith; that is, "they are defined or specified by a certain experimental intervention." And yet, although experimental means affect the form of the physical laws, the content of the mathematical equations which contain the laws do not derive from the particular experimental methods used. Here again the analogy between the *materia secunda* of physics and the geometric plane of Euclidean geometry becomes useful, because in the case of geometry also, while the manner of operation of a geometer affects the geometric properties of what is constructed, whether the geometer draws a square or a circle, the mathematical structures of the geometric forms are not determined by the geometer. They belong ultimately to the intelligible world which Smith quite rightly associates with the Platonic understanding of this term. This whole analysis is very much in accord with my views and opens the door for those who are able to understand the tenets of the sophia perennis as well as modern physics, for the integration of quantum mechanics into the traditional hierarchy of knowledge. It invites the integration of the world with which quantum mechanics deals into the ontological hierarchy of the perennial philosophy.

To achieve this end, it is of crucial importance to realize that the physical objects of quantum mechanics are not "things" in the ordinary sense but of much smaller dimension. Rather, they are relational and do not possess an *esse*. The main lesson to learn from this truth is that the essential attributes of things, therefore, come not from the quantum objects but from elsewhere or more precisely from above. Smith offers a wonderful metaphor by saying, "the essence of a plant, after all, derives from the seed, and not from the ground in which the seed is planted." One might say that the seed in this metaphor refers ultimately to the archetypal reality or Platonic Idea which the particular plant in question reflects and manifests on the corporeal plane. It is a valuable contribution by Smith to assert that modern physics does not deal with essence, that it is "inessential," because the realization of this truth provides an opportunity for those reductionists who are looking for the sun in the bottom of the well to cast their eyes above in order to see the origin of the essences which one observes and experiences in the corporeal world.

From my perspective, Professor Smith is completely right when he asserts that "the decisive step in the restitution of the cosmologia perennis is without question the rediscovery of 'forms' as an ontological and causal principle," and in fact on several occasions in my writings I have expressed the same idea. And it is also completely true that the destruction of the significance of forms based on incomprehension of their real significance by Bacon and Descartes opened the door for the reductionism of modern science and the constant attempt by scientists to explain things by their parts, asserting that the whole is no more than the sum of its parts. When Smith states, however, that this destruction of forms "has not... led towards the realization of the reductionist goal" and that "the reductionist philosophy appears also to have outlived its established usefulness as a heuristic principle," it seems to me that he is looking only at a few scientists like himself and not at the impact of reductionism associated with science on the general cultural scene of today. It is enough to look at the current mainstream view of the physical world, of medicine and the body, of the approach to the solution of social, economic, and ecological problems to see how entrenched the reductionist view really is. The philosophy of wholeness still remains in the margin of modern and postmodern man's worldview thanks mostly to mainstream modern science.

Furthermore, Smith makes the crucial statement that there exist in the world non-mathematical formal principles which are "none other than the aforementioned substantial forms, which prove moreover to be 'essential' in a strict ontological sense." He adds further that these forms constitute an ontological order. If this central statement of Smith's, with which I agree whole-heartedly, were to be accepted fully by modern science, a new scientific view would be born that would cease to be that of modern science as it is known today and become transformed into a further extension of the traditional sciences as I have already proposed. Then and only then could one say that reductionism has ceased to be operative and has outlived its "usefulness." Until then, unfortunately reductionism continues to level things to their lowest common denominator, to destroy quality in the name of quantity and to impoverish the spiritual vision and the minds of those affected by its siren call.

The essay of Wolfgang Smith, to which I have only added a few commentaries, is a seminal essay and should be studied carefully by all interested in the reintegration of science into the metaphysics contained in the heart of the sophia perennis and the traditional cosmological sciences associated with it. It should also be necessary reading for all searching for a new and richer philosophy for science, for those who often end with superficial adaptations of Taoism or Hinduism which are then related to the findings of modern science. Smith takes us much farther in this quest and shows the role that the sophia perennis can play in the veritable understanding of the significance of quantum mechanics and the integration of scientific knowledge into the universal hierarchy of knowledge.

For some forty years I have been writing on traditional science in relation to modern science, on Islamic science, and the hierarchy of knowledge as well as on traditional metaphysics and the perennial philosophy. It is an exhilarating experience for me to see here almost a synthesis of my own thought with many new insights on the subject presented by an active scientist also well versed in traditional doctrines. In writing these lines I feel as if I am expanding some of my thoughts in the direction of a new horizon opened by Smith. There is practically nothing in this important text to which I would need to respond in a critical manner in order to clarify differences with my own thought. On the contrary, to understand my thought on the subject of the relation between the sophia perennis and modern science, it is important to pay attention to my confirmation of the main theses of Smith's essay, which do full justice on the one hand to traditional doctrines and the tenets of the sophia perennis, to quote his terminology, and on the other to the discoveries of quantum mechanics and the nature of the whole venture of modern physics. I am happy that the occasion to write for this volume made it possible for Professor Smith to produce this exceptionally important essay in the field of the relation between the perennial philosophy and modern science, a field which has preoccupied me since my student days.

Seyyed Hossein Nasr
May 31, 2001

Chapter XIII
Wolfgang Smith, Traditionalist Catholic

For many years before I met Wolfgang Smith, I was well acquainted with his groundbreaking work. I had copies of all his books and articles, and I benefited from them greatly. In 2009, I wanted to ask if he would contribute an article to a book I was editing. Eventually, I was able to obtain his postal address from mutual friends, who told me that he did not use email, and the only way to contact him was by letter. His whereabouts were largely unknown, and scant personal details have been available about him on the internet. He lived with his wife as a recluse—in the world, but not of it, consistent with Christ's reminder that "My kingdom is not of this world" (John 18:36).

A few days after I wrote to him, I was honored to receive a telephone call from Wolfgang expressing his gratitude for my letter inviting him to submit a piece for the book. It was the first of many memorable conversations. Less than a month later, sometime around March 30, 2009, I was grateful to meet him and his wife, Thea, in person in Ojai, California, which is a little over thirty miles from Camarillo. His wife was an exceptional and talented person, who was very devout in her Catholic faith.

I remember as lunch was served at our table, both he and Thea expressed the need to give thanks for the food that we were about to receive, and so they started to pray over it. The patio of the restaurant was very crowded, yet they undertook their consecration without hesitation. It was a beautiful moment wherein the Divine presence was acknowledged in the here and now. We could not have asked for a lovelier day: it was a truly stunning spring afternoon. Our conversation, which spanned many themes—our sojourns to India, the state of the world, mutual friends, and essential books—always returned to our relationship with the Lord, as this was the *sine qua non* of the human condition. After this providential meeting, we had several more telephone conversations and written exchanges.

Wolfgang was born in 1930 in Vienna. Following the outbreak of the Second World War, he and his family fled Austria for the United States. He graduated from Cornell University at the age of eighteen with degrees in physics, mathematics, and philosophy. In very little time, he also obtained a master's degree in physics at Purdue University. He then earned a doctorate from Columbia University and served as a professor of mathematics at the Massachusetts Institute of Technology, the University of California (LA), and Oregon State University until his retirement in 1992.

He has expressed, on many occasions, how fortunate he was to have created his own mathematical theory of submersions so that he could co-exist but did not have to deal with the polemics within contemporary academia. While mathematics interested him and he had many significant publications in this field, he was more interested in the pursuit of science and metaphysics. Wolfgang discussed the profound isolation he experienced throughout his long tenure as a professor, as he never had anyone that he could call a friend among his colleagues. This left a permanent impression on him, not only about the state of academia, but also regarding the world around him.

Several years later, I was informed by mutual friends that his beloved wife Thea (Dorothy Marguerite Smith, born in 1926) had passed away at the age of 90 on June 21, 2017. I wrote to Wolfgang right away to offer my condolences. He phoned me shortly afterwards and we spoke for a long time about Thea and the significant role she played in his life. Indeed, it was Thea's influence that helped guide Wolfgang back to the Christian tradition. She had, in fact, spent some seventeen years in the Catholic convent of the Sisters of St Joseph of Carondelet, which she left after witnessing the horrors following the Second Vatican Council (1962–1965), according to the insistence of her spiritual advisor. She informed her husband that "God sent me to the convent in order to catch a big fish": this, of course, being a reference to none other than Wolfgang himself. We also discussed his new body of work, and the forthcoming full-length film devoted to his life and thought, *The End of Quantum Reality*, which was released in January 2020.

Wolfgang's decision to move forward and participate in this project was very much a surprise, given the film's personal nature and its intimate portrayal of his life and thought. I say this because I had known him to be a very private man who was averse to having the details of his life exposed for public consumption. His decision, however, was a profoundly compassionate offering that is sure to benefit many people. While I was conducting an interview with Wolfgang about this movie, we spent the entire day and most of the night together.[1] While we were discussing many important subjects, I recall, at one point, sitting in his lovely living room listening to *The Best of Serge Jaroff and His Don Cossacks*, appreciating the power and the brilliance of the voices. He had a great love of classical music and, when he could no longer attend concerts in person, he would keenly watch live performances from home.

On many occasions, Wolfgang expressed, in no uncertain terms, that if it were not for Rick DeLano (1957–2022), the producer of the film—who came into his life just as Thea was leaving this world—he himself would surely have also

[1] See Samuel Bendeck Sotillos, "The End of Quantum Reality: A Conversation with Wolfgang Smith," *Sacred Web: A Journal of Tradition and Modernity*, Vol. 45 (Summer 2020): 48–65.

died shortly after. The grief that he felt was a shattering blow to Wolfgang. It was very difficult for him to envisage a life without—they had been inseparable. His new work, along with the film, conferred great blessings and spiritual strength on Wolfgang, which allowed him to continue in the face of his terrible loss. Apart from her doctorate in mathematics, Thea was also a musician, dancer, and singer—in fact, she wrote a song conveying the great respect she felt for her beloved husband called "Lonely Man," which was featured in the film. Wolfgang provided the backstory to the song, saying that Thea had gone for a walk one day on their remote property of some twenty-five acres (outside of Corvallis, Oregon). Upon returning home, she told him that the song simply came to her and so she sang it for him.

Although Wolfgang saw metaphysics as something universal and timeless, he always returned to the uniqueness of the Christian tradition and stressed the need to pray constantly by remembering the "one thing needful" (Luke 10:42). He stressed that the Trinity and the Incarnation could not be found in any other religion. He was adamant about this, emphasizing that this unique feature could not be played down, for to do so was to misunderstand Christianity itself. Needless to say, he took to heart the Gospel teaching: "[T]his is eternal life, that they know Thee the only true God, and Jesus Christ whom Thou hast sent" (John 17:3). He felt blessed to have visited Padre Pio (1887–1968) in San Giovanni Rotondo, as he considered him as one of the greatest modern saints of the Catholic Church, and was also privileged to have participated in a public audience with Pope Pius XII (1876–1958) in Rome.

At the age of 18, he frequented meetings of the Vedanta Society in New York, under the tutelage of Swami Nikhilananda (1895–1973). He traveled extensively in India to deepen his acquaintance with the Hindu tradition, and spent time in the company of Śri Anandamayi Ma (1896–1982), one of the great spiritual luminaries of the twentieth century. Wolfgang was moved by the maxim of the Maharajas of Benares attesting that "There is no religion higher than Truth," which Nikhilananda wrote in a signed copy of his translation of the Upanishads. For Wolfgang, this confirmed a profound truth that remained with him until his final days. The Christian tradition confirms this powerful statement with Christ's dictum: "And ye shall know the truth, and the truth shall make you free" (John 8:32).

He spoke with deep disappointment about the death blow that the Second Vatican Council inflicted on the Catholic Church, and was devastated by the fact that the institutional church was on the brink of collapse. He spoke of an esoteric Christianity that was oral, rather than written, which allowed for the transmission of *gnosis* (transcendent wisdom) as expounded by Clement of Alexandria (c. 150–c. 215)—a faculty of spiritual wisdom unknown to most Christians today. He stressed that both written and oral traditions—comprising the exoteric and esoteric dimensions of the faith—were indispensable.

Wolfgang firmly believed it would soon be time for a new ecclesial body to emerge where the church would undergo a metamorphosis akin to the three stages of the life of Christ. Malachi Martin (1921–1999) referred to the second stage as the "Church of the Entombment" (or underground church), which we are currently in, and called its final stage the "Church of the Resurrection." He often spoke about Malachi Martin, whom he knew personally, and whose work he considered to be profoundly significant. Wolfgang also took the view that the mission of this new church, along with its Mystical Body, could only be fulfilled when Roman Catholicism and Eastern Orthodoxy (and the Protestant forms of Christianity as well) were reunited, as both were needed to complete one another—"the Church must breathe with her two lungs" as St John Paul II once remarked. Only then could there be a truly *catholicos* or universal Church. This new church would barely resemble how Catholicism is viewed now, because the former would be more mystical in its outlook. Wolfgang recalled that he and Father Martin agreed that the Second Vatican Council is an evil in response to which it is necessary for the Church of the Resurrection to be realized.[2]

Wolfgang made no attempt to gloss over the dark episodes in the history of the Catholic Church. He spoke despairingly of the Inquisition, stressing that it was not only "dreadful" but a "terrible crime!" He often affirmed the motto "The corruption of the best is the worst" (*corruptio optimi pessima*). In this regard, we recall the following passage from St Paul: "For we wrestle not against flesh and blood, but against principalities, against powers, against the rulers of darkness of this world, against spiritual wickedness in high places" (Ephesians 6:12).

I remember having a long conversation with Wolfgang on February 18, 2020—on the occasion of his 90th birthday—and observing how assiduously he was working on his new projects. He demonstrated remarkable energy and commitment in tirelessly pursuing this important work, noting that if he were to stop writing, he would likely breathe his last before too long. Mind you, when I initially met Wolfgang years earlier, he explicitly stated that he was putting down his pen and retiring from writing. Many would be grateful that he did not follow through on this resolution, and it is truly remarkable to see how much more he has accomplished since then.

I once asked how he wished to be remembered, to which he replied: "As a believing Christian." When asked how he would describe his contribution to science, he said: "When we are born we have a destiny, a way that we are meant to go; to be true to this or not. My interest in physics and mathematics was something that I was supposed to do. To follow this path—as determined by God—is more beautiful than anyone can imagine." He again repeated that his task was "to fulfill my destiny." Wolfgang proved, once and for all,

[2] See *In Quest of Catholicity: Malachi Martin Responds to Wolfgang Smith* (Brooklyn, NY: Angelico Press, 2016).

that science disconnected from metaphysics led, not only to an acute imbalance, but to the nefarious ideology of scientism, along with the pseudo-religion of secularized, postmodern humanity. For him, science could only be fully integrated when it returned to its spiritual roots in the transcendent.

In several of our conversations, Wolfgang indicated that he was not sure how long he had left to live, and he had come very close to dying on several occasions (in the Nazi invasion during the Second World War, and on several mountain-climbing expeditions). He valiantly fought through these trials so that he could continue his work. On one of our visits, Wolfgang appeared to be struggling very much with his health. He could barely draw breath, and walking was very difficult for him. I was unsure how long he could endure this condition. Yet, there were moments was he was so vibrant, present, and articulate, not to mention steadfast in his demeanor, that it reminded me of the Wolfgang I had known years before. Regarding what lies beyond the grave, he said that our post-mortem states were an imponderable enigma, about which he declined to speculate.

During the recent pandemic, we spoke frequently over the phone, due to the shelter-in-place order that forced him into severe isolation. He would always say that "this too shall pass," insisting that we can never give up hope or fall into despair. He observed that, in our most difficult moments, grace is even more abundant—our primary duty, then, is to always seek refuge in God and to pray continually. Throughout this time, Wolfgang told me that there were days when he would barely eat, as he was working almost around the clock, which allowed him to uncover new ground and deepen his insights. So, although this was a very testing time, it was also a blessing because it provided him with the opportunity to bring to a completion his life's *opus*. During this uncertain period, he would often bring to mind the following verse: "He that is not with me is against me" (Matthew 12:30), indicating that we all have a choice to make in this earthly existence and that, unless we conform our will to God's will, we are—knowingly or unknowingly—working against Him. Wolfgang never failed to place his trust in God, saying it was imperative that we live a prayerful and contemplative life.

During the COVID crisis, he felt the need to partake in the Eucharist regularly, to keep him "in a state of grace," and he reflected on how ominous it was that God's churches were closed down due to the lockdown, especially at a time when people most needed access to them. His prayer was heard, as he was able to find a priest who invited him—in secret—to attend Mass during the pandemic. He was truly grateful for this and celebrated these occasions with great joy. When the churches finally reopened, he lamented that Holy Communion could no longer be given directly on the tongue; he considered Communion on the hand to be a diabolical subversion.

Wolfgang noted that the forces of darkness were very powerful in our time and that, while these liturgical restrictions were touted as a public safety measure, they were nonetheless a pretext to undermine the traditional Catholic Church. Through an underground network, he was able to make contact with other priests who were also offering the traditional Mass in the correct manner (albeit clandestinely). All the while, he never forgot these comforting words: "I am with you always, even unto the end of the world" (Matthew 28:20). Needless to say, he gave thanks when he was able to return to St Mary Magdalen's Chapel, where he could once again attend the Tridentine Mass and receive the Sacred Host on the tongue as he was previously able to do.

He imparted the advice, on several occasions, that when it comes to our spiritual life, we need to "read selectively," which was an important lesson for him to learn as, early in his life, he read everything of interest that he could get his hands on. He added that "one will drown no matter how far they can swim; you cannot reach the shore because there is no way to assimilate it all." For this reason, he had learned—through the course of his long and blessed life—that what was of utmost importance was to focus our spiritual life on Christ, in order to return to God.

In March of 2023, Wolfgang came close to dying when he took a very bad fall down the steps of his two-story home, hitting his head and breaking several ribs. Although he was able to do some writing following the accident (in order to complete the publication of his collected works through the Philos-Sophia Initiative), he told me many times that he had not fully recovered from this incident. On Christmas Eve of 2023, he suffered another serious fall, which then led him to be placed in an assisted living facility to support his increased needs. It was at the end of May 2024 that I was able to visit Wolfgang for the last time prior to his repose in his Lord.

Wolfgang was exceptionally serious in his demeanor, and had a razor-sharp intellect. He was also a very eloquent wordsmith, always being mindful to communicate in a precise manner by using words correctly. He was—without fail—the perfect gentleman, extending graciousness and generosity towards everyone. Wolfgang was a man of another age, as was Thea: an age that has been long forgotten by the world today. He always felt blessed to have lived as long as he did, and believed that this could not have been possible unless God had willed it. For his tombstone, Wolfgang selected the following inscription: *Deus Caritas Est* (God is Love), for all things are embraced in the Divine mercy. This is taken from the verse "God is love; and he that dwelleth in love dwelleth in God, and God in him" (1 John 4:16). He insisted that *love* was not just a sentimental notion, but a way of life.

Samuel Bendeck Sotillos

Chapter XIV

Against Teilhard de Chardin's Theology-Fiction

Pasolini once wrote that theology is one of the branches of fantasy literature. We will reserve this assertion for the fictional theology of Pierre Teilhard de Chardin (1881–1955). Despite eleven condemnations by the Church, which silenced him because of the "*serious attacks on Catholic doctrine*" developed in his books, he enjoyed incredible success in the 1950s and especially in the 1960s. Even good minds were seduced by his lyrical prose filled with neologisms and attractive, albeit bizarre, poetic flights of fancy. His euphoric project of reconciling modern science and faith was very much in keeping with the spirit of a time bathed in optimism against a backdrop of technological advances. The great theologian Henri de Lubac was commissioned by the Society of Jesus to defend his colleague. Previously, Teilhard had submitted the manuscript of *The Human Phenomenon* to him: Lubac sent him back 240 requests for corrections!

Even though most Christians today are unaware of the work of the paleontologist-theologian, the Jesuits are trying to keep the flame alive. The Loyola Paris Faculties (formerly Center Sèvres) have a chair named after him. Courses and symposiums are organized from time to time. In 2023, a Teilhard de Chardin Center was opened in the heart of the Saclay technology and science hub, supported by the Society of Jesus and the dioceses of the Paris region, with the ambition of making it "*a space for dialogue between science, philosophy, and spirituality.*" However, not only are the Jesuit's theses devoid of scientific credibility, but they are also contrary to the most basic truths of faith. This has been brilliantly demonstrated by Wolfgang Smith in his book *L'évolutionnisme théiste de Teilhard de Chardin*[1] (*The Theistic Evolutionism of Teilhard de Chardin*), drawing on his dual expertise as an inspired philosopher and a high-level scientist, physicist, and mathematician. Teilhard was a paleontologist without any particular genius, incompetent in both biology and physics, and the knowledge he gained from his profession (the discovery of fossils) had no connection with his ideological construct, as he himself acknowledged in his correspondence. This explains why his theses have been criticized by renowned scholars.

His intention was to reintroduce God into a scientific vision dominated by evolution. He made no secret of his intention to rebuild Christianity—an "*improved Christianity,*" an "*ultra-Christianity,*" a "*meta-Christianity,*" as he put it—on new foundations. The universe was conceived in a pantheistic mode: "*There is neither Spirit nor Matter in the World; the Fabric of the Universe is Spirit-Matter.*" According to a great "*Law of Complexity/Consciousness,*" "*Everything that exists is Matter becoming Spirit.*" For him, the doctrine of creation *ex nihilo* is contradicted by "*temporal-spatial enormity,*" "*energetic immensities,*" and the "*unfathomable organic connections of the phenomenal world.*" For Teilhard, "*grace represents a physical super-creation. In other words, it is of a strictly biological nature.*"

We must also abandon our conception of a God above time: "*Around us and within us, through the encounter of his Attraction and our Thought, God is in the process of 'changing' ... As the quantity of cosmic union increases, its brilliance and hue are enriched!*" Rejecting the law of increasing entropy in the universe, Teilhard asserts that everything is irresistibly converging toward an "*Omega Point,*" which is none other than the cosmic Christ.

[1] L'Harmattan, 2023.

One of the stumbling blocks between Teilhard's vision and the teaching of the Church is the dogma of original sin. In what is perhaps the most fascinating chapter of his book, Wolfgang Smith explains how Adam's Fall fits into a different representation of origins than that of the Jesuit: "*it is this primordial catastrophe, and not Darwinian evolution, that is responsible for the condition we know today.*" On the contrary, for Teilhard, evil is simply comparable to the disorder caused by natural processes. This conception leads him to a surprising relativization of human sin. His "*neo-humanist*" political vision was based on the phenomenon of "*socialization*," aggregation through collectivization, which led him to a singular complacency toward Nazi and Communist totalitarianism. In 1938, he wrote: "*I don't know where to place my sympathies at the present time: where is there more hope and idealism at the moment? In Russia, or in Berlin?*"

All truths of faith are ultimately reinterpreted in his own way, for better or worse; or, in the case of the most troublesome ones, abandoned.

The publication of Wolfgang Smith's essay reignited vigorous controversy in France. Teilhardians (often very elderly) deplored—without always reading it—the criticism of a thinker who had inspired their thinking, while others recalled the terms of the Monitum of the Holy Office of June 30, 1962: "*Certain works by Fr. Pierre Teilhard de Chardin, also published after his death, are spreading and enjoying great success. Without passing judgment on matters relating to the positive sciences, it is quite clear that, on the philosophical and theological level, these works are full of ambiguities and even serious errors that offend Catholic doctrine.*" The strength of Wolfgang Smith's analysis lies in his deconstruction of the core of Teilhard's fictional theology.

Denis Sureau

Chapter XV
Putting an End to Evolutionism

In a work that from the outset displays a desire to defend the Christian perspective on the meaning of man's presence in the universe, certain subjects occupy a privileged place in Wolfgang Smith's writing and, approached from a scientific, philosophical, and religious point of view, are the subject of repeated developments in most of his writings. Smith did not hesitate to insist on the need to return as often as necessary to the dangers of a certain ideology skillfully maintained in scientific circles, whatever the reasons for it, and to denounce it all the more vigorously as his own advances in the field of physics provided him, over the years, with new arguments in favor of his own convictions. Such is the case with evolutionism, a theory as tenacious as it is devoid of any solid scientific basis, but the subject of a veritable secular religion, maintained for nearly 150 years and now rooted in the mentality of the ill-informed man in the street as much as in that of theologians who have managed, by some feat of magic, to integrate it into a Christian vision of creation.

This is why, inspired by a constant search for reality and truth, Wolfgang Smith made a significant contribution to an anti-evolutionist school of thought that, in the 1950s, already included renowned scientists such as Maurice Vernet, Louis Bounoure, Jean Rostand, the latter having confided to *Le Figaro*, at the height of the great evolutionary fever, these distressed remarks that seem to have been misunderstood: "The world postulated by transformism is a fairy-tale world, phantasmagorical, surreal. The important point is that we have never witnessed, even on a small scale, a genuine phenomenon of evolution." He concluded that despite its "indigestible" nature, he was obliged to believe in it "because he could see no other way."[1] Such is the pressure of scientific consensus in certain cases.

Darwin's theory is too well known for us to feel it necessary to define it. Let us say, however, at the outset that Wolfgang Smith, when discussing it, clearly states that the notion of transformism (macroevolution) has already been ruled out, and that he is interested exclusively in what has been called microevolution, the numerous and slight changes that occur over prolonged periods. And we quickly discover that if this form of evolutionism occupies an important place in Smith's work, it is because this theory has had a considerable impact on a question that is particularly close to his heart, namely *the concept of the Divine in modern man and the place of tradition in today's world.* Do we know whether Darwin was willing to consider divine intervention in creation? Although he was fairly discreet on this subject, we know from his notebooks that he ruled out the possibility. On the other hand, we also know that many evolutionists, past and present, have been and are resolutely materialistic. This is evidenced, among other things, by the very explicit statement of a prominent evolutionist, Richard Lewontin, professing that "[his] *a priori* attachment to material causes is absolute, for we cannot tolerate the presence of a divine foot in the door."[2] And, in the course of its development, Darwinism has clearly taken on an anti-religious character.

In the opening chapter of his first book, *Cosmos & Transcendence, Breaking Through the Barrier of Scientistic Belief,*[3] addressing the theme of evolutionism, Smith notes, almost *in passing,* "a remarkable theory [that] deserves mention": it is an explanation that resolves in its own way the problem of the lack of tangible evidence of continuity from one animal phylum to another, regarding which French paleontologist

[1] *Le Figaro Littéraire,* April 20, 1957.

[2] Ibid.

[3] *Cosmos & Transcendence,* 69, n.10.

Louis Bounoure observes "one must be touched by evolutionary grace to find this reasoning convincing."[4] Clearly, Wolfgang Smith was not touched by this grace, and on the contrary, he never ceased to oppose the author of this astonishing discovery, the Jesuit priest Pierre Teilhard de Chardin, whose written and published work is in flagrant contradiction with Smith's vision of global Reality. In fact, when it came to evolutionism, Smith, returning to the subject in most of his writings, always accompanied it with a detailed account of what he believed to be the origin and development of life in our universe as he conceived it. Taking advantage of the criticisms of renowned scientists opposed to this theory, he also draws his argument from the philosophical and metaphysical principles of what he calls the *philosophia perennis*. This is where his originality lies in relation to his contemporaries opposed to Darwinism, who generally invoke only scientific reasons. Smith revisits the subject in virtually all his works, but not before vilifying its expression in all its aspects by its official prophet (Teilhard de Chardin, glimpsed in *Cosmos & Transcendence*), to whom he devotes his second *opus: Teilhardism, the New Religion.*[5]

Before continuing our examination of Wolfgang Smith's contribution to the subject, we must return to Darwinism and its turbulent history. Darwin's theory is ultimately based on the evidence (and law) of survival of the fittest. In its early days, and until the middle of the last century, however improbable it may have been, it was still scientifically conceivable, as knowledge of biology was insufficient at the time to seriously challenge it. It was only with the discovery of the genome and DNA that it began to falter. However, from the outset, Darwin struggled to provide convincing arguments. In his search for tangible evidence and relevant intermediate forms, he came

[4] At the Congress of Philosophy of Science held in Paris in 1949, quoted in *Cosmos & Transcendence*, 69, n.10.

[5] *Teilhardism and the New Religion* (Rockford, IL: TAN Books and Publishers, Inc, 1988).

up against what is known in paleontology as "fossil stasis" and in biology as the absence of "vestigial organs" (some that were thought to have been discovered would later be recognized as having a definite function), as well as "nascent organs" in the process of formation. He was thus forced to note the prolonged existence for centuries of stable genera, species, and families, showing no change... before suddenly, completely new and unknown lineages appeared; all this long before advances in biology finally categorically refuted the claims of evolutionism and its dogmatism. On this last point, the prize must go to Haeckel (author of the famous "tree of species"), when he declares: "Nothing could be more absurd, nothing could more clearly demonstrate a lack of intelligence in our theory than *the requirement that it be based on empirical evidence*."[6] All this is a far cry from Claude Bernard's dictum: "Our theories should only be believed on the basis of experimental evidence."[7]

From the 1950s onwards, things began to change and accelerate. In 1953, John Francis Crick discovered DNA, the molecule that contains our genetic makeup. But at around the same time, certain philosophers began to explore a new concept, the *anthropic principle*. Indeed, if the universe we inhabit is an assembly of irreducible totalities, which themselves can be traced back to a single irreducible Wholeness belonging to the eternal realm, according to the tripartite plan presented by Wolfgang Smith (see previous chapters of this book), we may ask ourselves about the meaning of this upward movement of life spread by this Totality, returning to Leibniz's question: "Why is there something rather than nothing?" and wonder what is the purpose of this almost uninterrupted and increasingly complex process of development? The Darwinists' answer, based on the theory of evolution according to the law of survival of the fittest, does not highlight any necessity for this upward movement other than a paradise finally achieved

[6] Quoted in *The Human Phenomenon*, 151, and *ÉT*, 41, n.18.

[7] Cited by Louis Bounoure, *Revue des Sciences Religieuses*, 1957.

on a physical level, the ultimate success of humanity achieving "triumphant humanism," even if it comes at the price of escaping to other planets, given the fate that awaits this one. The question of "meaning" arises, and the "necessity" of a first cause and a finality remains.

Since the publication of *The Anthropological Principle* by John Barrow and Frank Tipler in 1962,[8] much water has passed under the bridge of the anthropic principle—weak or strong version—proposed by the authors; in other words, the principle of a creative intelligence providing a habitable environment for a humanity destined to emerge (incidentally, this is the opposite of the order of creation according to the Bible, in which Man preceded the Garden). This principle has sparked much controversy and exchanges of more or less veiled insults. Since then, scientific progress has provoked confrontations of arguments that are almost as tense, but more meaningful. In 1982, Paul Davies published his book, which "asks why, among the infinite range of values that Nature could have selected as fundamental constants, and among the infinite variety of initial conditions that could have characterized the early universe, the values and conditions that prevailed produce the range of particular features that we observe."[9] Last but not least, Wolfgang Smith tells us that in 1996, molecular biologist Michael Behe, in a treatise entitled *Darwin's Black Box,* introduced the concept of "*irreducible complexity*" and convincingly showed "that no natural process can give rise to '*irreducibly complex*' structures."[10] His treatise was widely circulated and debated in scientific circles. The essence of his theory can be summed up in the following definition: "By *irreducibly complex*, I mean a single system

[8] John Barrow and Franck Tippler, *The Anthropic Principle* (Oxford University Press, 1962).

[9] *The Accidental Universe* (Cambridge University Press, 1982), 11. Quoted in *ÉT,* 45.

[10] *Sagesse de la cosmologie ancienne: Les cosmologies traditionnelles face à la science contemporaine* (hereafter *SCA*) (L'Harmattan, 2008), 221.

composed of several well-fitted parts that contribute to a basic function, and where the removal of one of these parts prevents the entire system from continuing to function."[11] The definition is intentionally formulated with the ulterior motive of ensuring that no Darwinian process has the slightest possibility of producing an irreducibly complex structure. Darwin himself observed that "if it could be demonstrated that any complex organism existed, which could not possibly have been formed by numerous, successive, slight modifications, [his] theory would absolutely break down"; to which he added: "but I can see no such case."[12] One of the most frequently cited examples of irreducible complexity is the impressive case of the bacterial flagellum, whose function is to propel bacteria in their aquatic environment, "a technological feat that defies imagination"[13] and which no Darwinian scenario could account for. But that's not all: let's hear the rest, as Smith himself reports it:

> Two years after Behe, in 1998, William Dembski, mathematician and philosopher, wondered whether it was possible to recognize that there is a purpose thanks to some signature or criterion that could be defined in strictly mathematical terms. It turns out that the theory that results in his treatise entitled *The Design Inference*[14] does not merely generalize Behe's concept of "irreducible complexity," but raises the question of deducing intelligent design *on a mathematical* and therefore rigorous *basis*. Dembski discovered that the signature or criterion of this design can actually be expressed in terms of a probabilistic notion of "complex specificity" or, more precisely, in terms of information theory, a concept *of complex*

[11] Ibid., 222.

[12] Ibid.

[13] Ibid., 223.

[14] *No Free Lunch* (New York: Rowman & Littlefield, 2002). See *SCA*, 227.

> *specified information* or CSI. The essential result is a conservation theorem for CSI, which states that CSI cannot be generated by any natural process, whether deterministic, random, or any combination of the two, as in so-called evolutionary algorithms.[15]

It is not surprising that Dembski has attracted criticism from various sectors of the establishment, above all from Darwinists. But ultimately, says Wolfgang Smith, "it is difficult to quarrel with a mathematical theory."[16]

Thus was born the theory of *intelligent design*, and the doctrine of evolution now finds itself seriously shaken. But could it be too late?

Indeed, around the middle of the century, a mind "touched by evolutionary grace" appeared, believing himself called to overhaul the Catholic religion in line with the evolutionary perspective, and soon took center stage. Drawing on his professional credentials as a paleontologist, Father Teilhard de Chardin did not hesitate to make assertions that were as peremptory as they were risky, for example: "Things are changing. Since the time of Darwin and Lamarck, numerous discoveries have established the existence of transitional forms postulated by the theory of evolution,"[17] an opinion in complete contradiction with what we mentioned earlier. The persuasive enthusiasm with which Teilhard propagated Darwin's theory was commensurate with his own convictions:

> A theory, a system, evolution? No; but much more than that, a general condition to which all theories, all hypotheses, all systems must now conform and satisfy in order to be conceivable and true. A light illuminating all facts, a curve that all lines must follow. That is what evolution is.[18]

15 *SCA*, 221–27.
16 Ibid., 222.
17 *The Human Phenomenon,* 83, *ÉT*, 40, n.5.
18 Ibid., 242, *ÉT*, 39.

These remarks led Hans Urs von Balthasar to say, "For Teilhard, evolution is the only category of thought."[19]

Let us note the expression: "*all* theories, *all* hypotheses, *all* systems." Thus, the origin of Man:

> Today, the evolutionary origin of Man is no longer in doubt for science. [...] Let it be clearly stated: the question has already been settled—so thoroughly settled that continuing to discuss it in schools is just as much a waste of time as if we were still deliberating on the impossibility of the Earth rotating.[20]

We should appreciate the incredible confidence with which such stunning words are uttered: "This *evidence* is now beyond verification, *as well as beyond any subsequent refutation by experiment*,"[21] as if Teilhard were unaware that, in order to be considered valid, a scientific proposition must be considered "falsifiable." Similarly, this astonishing assertion appears in *The Heart of Matter*, a work composed a month before his death:

> In truth, Christ saves
> —but should we not immediately add that he is *at the same time saved by Evolution*?[22]

which invites us to address the thorny issue of evolutionism as it manifests itself in God, according to Teilhard, and which Smith tackles head-on and in detail in a chapter of his book on *Teilhard de Chardin's theistic evolutionism.*[23]

In the chapter entitled "The God of Evolution," Smith notes that "the new theology had to revolve around the incarnate Christ, but naturally, the figure of Christ underwent a

[19] "Die Spiritualität Teilhards de Chardin," *Wort und Wahrheit*, XVII, (1963), 144. Quoted in *ÉT*, 307.

[20] *Comment je crois*, 164, quoted in *ÉT*, 43.

[21] Ibid., 151, *ET*, 39.

[22] *The Heart of Matter*, 186, *ÉT*, 164.

[23] *ÉT*, chap. VI, 143–78.

change. He was transformed into a kind of cosmic Christ, clothed in the body of the universe."[24] Above all, he became "an *evolving* Christ." One wonders, then, how this cosmogonic Christ relates to God the Father. What remains of an *ad intra* theology of the Trinitarian doctrine? In *The Human Phenomenon,* Teilhard teaches us that "if the World is convergent, and if Christ occupies its center, then the Christogenesis of St Paul and St John is nothing less than the extension of Noogenesis, in which Cosmogenesis culminates."[25] However, it should be remembered that there is not *one* Christogenesis of St Paul, but two: first, the eternal procession of the Son, the Word who was "in the beginning," and second, the human birth of Christ "who became flesh and dwelt among us" (Jn 1:14). To which Christogenesis is Teilhard referring? Only a cosmic Christ could possibly be considered an extension of the natural evolutionary process. But then, what about the *ad intra* doctrine of the Logos? Does Teilhard accept the idea of the eternally begotten Word? According to his reasoning, the recognition (by St John) of the Alexandrian Logos was a crucial and certainly inspired step, but it must now be followed by permission, in fact *the command, from the Holy Spirit* to use the same tactic (*scientific* means) in order to conceive a better and definitive formulation of Christian truth. Teilhard is the new apostle, authorized to identify Christ, not with the Alexandrian Logos, but with the "neo-Logos of modern philosophy, the principal evolver of a universe in motion."

It is also interesting to consider what an evolutionary theology has to say on the subject of a God conceived as immutable, as he appears in the words of the Bible and the Gospels. "Creatures change, but you remain," says the psalmist (Ps. 101:27–28). As the author of time, God is not affected

24 Ibid., 164.

25 *The Human Phenomenon,* 330–31, *ÉT*, 165.

by time: "Before Abraham was, *I am*"—a truth vital to the Christian faith, and St Augustine emphasizes the link between this phrase spoken in John 8:58 and the formidable warning in John 8:24: "If you do not believe that 'I am,' you will die in your sin." At the end of his life, Teilhard, speaking of his transitional period of relative orthodoxy, wrote:

> I did not realize that inevitably, as God "metamorphosed" the world from the depths of Matter to the heights of Spirit, the world in turn had to "metamorphose" God. Under the very effect of the unitive operation that reveals Him to us, God, in a way, *transforms Himself by incorporating us.*[26]

It matters little to Teilhard that Christ Himself declared, "Be perfect as your heavenly Father *is* perfect." God depends on evolution for "His completion and fulfillment," which Smith cites ironically, concluding, "As if this fact alone were not enough to make the 'Absolute Being' less than absolute!"[27]

What is striking here is that Teilhard seems to ignore the reality of a divine Essence in which the entire creation has been grounded since "the beginning." However difficult it may be for an inhabitant of the earth to envisage this presence of the temporal in the eternal, the fact remains that the evolutionary unfolding of the universe and its completion in God is inscribed in the "Alexandrian" Logos and needs no completion or modification. "Man is a being who has been commanded to become God," says St Bonaventure. And the Gospel shows him the way to go. Teilhard's contortions add nothing to their tortuous demonstration of a so-called new path. In this regard, a work such as that written during his long life by our contemporary Wolfgang Smith, a high-level scientist, philosopher, and metaphysician, teaches us much more.

To return to evolutionism and conclude on the subject, it is clear that its history, from its emergence to the first

[26] *The Heart of Matter,* 64, *ÉT,* 147.
[27] *ÉT,* 148.

decades of the twenty-first century, amounts to the chronicle of a death foretold. But on the one hand, philosopher Thomas Kuhn[28] pointed out that for a theory to disappear and for there to be a "change of algorithm," it must be replaced by a theory that is considered better and more consistent with reality. In other words, fallacious theories die hard. And on the other hand, it so happens that an unexpected avatar of this doctrine appeared at the moment when its decline was beginning, in the work of a fundamentally religious mind gifted with boundless imagination and an extraordinary love of "holy Matter," who, synthesizing authentic faith and his "evolutionary obsession" (according to Jean Borella), was, in the mid-twentieth century, the instigator of a "*theistic* evolutionism," which quickly won over the best minds and became a tidal wave among the simple. This "New Arius" (as Wolfgang Smith calls him), while creating a rebound in the doctrine of biological evolutionism, gave it an unsuspected scope by extending it to the human psyche, society, and the bright future of the cosmos based on the unsuspected progress of science, and above all religion, thanks to the "*endomorphization*" of the Divine in the life of the Earth. This naturally necessitated serious changes to the "*old Church*," whose "*foundations*" had to *be* "*shifted*"—which are and always will be Christ, according to St Paul—and which had to be "*freed from the Neolithic God*" to whom it was then enslaved. Let us reassure ourselves: "*it would still be Christianity, of course!*"

It is not our place here to develop in detail the many aspects of Teilhard's program, which is the subject of a presentation in this book, but to note that Kuhn's remark quoted above invites us not to waver in our determination to make Truth triumph, arming ourselves with a sense of Reality. This is something that did not escape Wolfgang Smith, who had set

[28] Thomas Kuhn, *The Structure of Scientific Revolutions* (University of Chicago Press, 1962).

himself this mission and emphasized in one of his prefaces that the obligation to constantly return to certain notions, even at the risk of repeating himself, was not a concern for him; which he did, to our great benefit.

Marie-José Jolivet

As a Conclusion

The End of Quantum Reality

A Conversation with Wolfgang Smith[1]

In January 2020, a full-length film was released on the life and work of Professor Smith, entitled The End of Quantum Reality *(Producer: Richard DeLano [1957–2022]; Director: KTEE Thomas). This interview, which was conducted at Wolfgang Smith's home in Camarillo, California on November 29, 2019, focuses on the long-awaited release of this epochal film and Professor Smith's legacy.*

Samuel Bendeck Sotillos (SBS): If you were asked to give a brief synopsis of your film, *The End of Quantum Reality*, how would you articulate this for audiences unfamiliar with your work?

Wolfgang Smith (WS): What we hope to accomplish is to deliver the audience from an erroneous worldview imposed upon us in the name of science, and in so doing deliver them from a chronic state of schizophrenia, which is a consequence of this worldview.

Physics has long claimed—on supposedly solid scientific grounds—that all things are simply composites of so-called fundamental particles. What we have proved—on the basis of quantum theory itself—is that such is not in fact the case: the world in which we live, and move, and have our being does *not* in fact reduce to mere particles! This means that the world we normally perceive by means of our five senses is not

[1] This interview first appeared in *Sacred Web: A Journal of Tradition and Modernity*, Vol. 45 (Summer 2020): 48–65. Reprinted by permission.

after all illusory, as we have been taught to believe since the Enlightenment.[2]

SBS: You have accomplished a remarkable feat in exposing the fundamental errors of contemporary science in an unparalleled fashion. Your findings fundamentally challenge the presiding ideology of our times and are considered "heretical" because they deconstruct the ideology as idolatry. How did this insight or process occur?

WS: First of all, I should say that, due to the circumstances of my life, I had an advantage over my colleagues. For in consequence of extended travel in remote parts of the globe, particularly in the Himalayan regions of India and Nepal, I had contact with living spiritual traditions which enabled me to break out of the limited *Weltanschauung* of our day. I learned to look at the world in a very different way. I thus enjoy the incomparable advantage of having been, to some degree at least, set free from the prevailing *Zeitgeist* which so stringently determines the way we view the world. I think this may well be what enabled me to arrive at certain insights which others, perhaps far better informed than I, were unable to perceive.

SBS: You speak of our time as "a singular moment in history" and make a connection between the "quantum reality problem" and a four-century arc of history René Guénon [1886–1951] has termed "the reign of quantity,"[3] which appears to be reaching its end. Can you please expand upon this important idea and discuss the relevance of the French metaphysician René Guénon upon your work?

[2] Editor: Later in this interview, Professor Smith elaborates on the ontological discrepancy between the *corporeal* and the *physical* realms, pointing to the necessity for a realignment of physics with traditional metaphysics.

[3] See René Guénon, *The Reign of Quantity and the Signs of the Times*, trans. Lord Northbourne (Hillsdale, NY: Sophia Perennis, 2004).

WS: It was really René Guénon who made me realize how strange our contemporary *Weltanschauung* is, in that it pretends to reduce all objective reality to the category of *quantity*. Guénon made me realize that *quantities* are correlative to *qualities*, and that whereas quantities derive from below, metaphysically speaking—that is to say, from the side of *materia*—qualities derive from above: from the side of *morphē* or *form* in the Aristotelian and Platonist sense. So that was a major rectification that proved to be a *sine qua non* for unraveling the enigma of the quantum world.

I first encountered the legacy of René Guénon in a Hindu monastery: I happened to wander through the library and came upon one of his books. I became very interested. Upon my return to the West, I made it a point to acquire his books, and there ensued a period, lasting more than a year, during which I was absolutely immersed in the study of Guénon's writings. And that has undoubtedly had a profound impact upon my thought. Guénon helped me to grasp what authentic metaphysics is about. In the modern West we have in effect forgotten what metaphysics is, due to the impact of philosophers like Immanuel Kant and his successors. We have forgotten what philosophy in the true sense—and metaphysics above all—is by right. René Guénon was a master in that regard. He has not only introduced me to what might be termed the perennial metaphysics, but has enabled me then to look on contemporary civilization from an authentic metaphysical point of vantage. And this is really the starting point of my serious work. Being enabled thus to look upon contemporary civilization from this point of vantage gave me a decisive advantage over other people concerned with these issues. So I am deeply grateful to René Guénon for this crucial inspiration.

SBS: You make a remarkable point that will come as a great surprise to many: namely, that there is very little empirical evidence for Einsteinian physics. You specifically refer to his theory of special relativity, and people are generally unaware that his formula $E = mc^2$ has nothing to do with the theory of relativity. Nowadays Einstein is regarded as one of the greatest minds of the twentieth century; and as you have yourself perceptively noted, the world itself has in a sense become "Einsteinian." The idea that he could have been wrong is virtually inconceivable to the contemporary mindset. Can you please explain?

WS: Well, I am glad you raised this question, as it touches upon a very important issue. I must admit that until a few years ago, I was among those who regarded the physics of Albert Einstein—in its main affirmations, at least—as sacrosanct. I was amazed therefore to find that this proves not to be the case. In the first place, as you have mentioned, it turns out that the famous formula $E = mc^2$, which Einstein derived from his special theory of relativity, is in fact implicit in the classical Newton-Maxwell physics, and actually appeared in the journal literature prior to the advent of Einsteinian relativity: Einstein himself admitted the fact in a book published in 1950[4]—but by then the harm had been done: the whole world, just about, had been converted to Einsteinian relativity largely on the strength of this fateful formula. Moreover, when I began thus to look more critically into Einsteinian physics, I discovered—to my great surprise—that there were experiments, dating back to 1913, that appeared to falsify the theory, while it turns out that experimental findings said to support it proved, on closer scrutiny, to be problematic. It appears there is not a single experimental result which unequivocally substantiates Einsteinian relativity.

For example, if you take the famous 1918 solar eclipse observations which are said to have corroborated the relativistic bending of light rays in the vicinity of the Sun, suffice it to say that these results are very far from a clear-cut verification

[4] *Out of My Later Years* (New York, NY: Philosophical Library, 1950).

of the general theory. For one thing, there are other possible explanations for the observed deflections: for example, the presence of an atmosphere around the Sun. Moreover, the numerical data appears not to match up that accurately with Einstein's predictions. On closer examination, thus, it appears that Einsteinian physics has not in truth been adequately vetted.

What is more, over the past decades a great number of predictions based upon Einstein's theory have actually proved false: the annals of particle physics, in particular, bear witness to this fact. Yet the most striking example of failure is the Einsteinian astrophysics, the celebrated Big Bang theory, which has been taught in our schools as the "scientific" account of how the world came to be. Well—unbelievable as it may seem to the thus indoctrinated mind—the scientific fact is that measurements of the CMB or Cosmic Microwave Background, obtained by the Planck satellite launched in 2009, have definitely disproved this Big Bang scenario, without so much as a possibility of redress. It would take us too far afield to go into the particulars. Suffice it to say that the Planck data has revealed the existence of a global axis in the CMB, which flatly contradicts the very assumptions of Einsteinian astrophysics. My point is that once you step out from the ranks of Einsteinian believers to take a critical look, the cracks in the Einsteinian edifice become readily apparent. For my part, I am persuaded the theory has in fact been disproved, and that its effect on physics as a whole has been disastrous.

SBS: A central dogma of the modern world is evolutionism. Could you please speak to this dogma in both its Darwinian form, and what has come to be known as "theistic" evolution?

WS: Regarding Darwinian evolution, the fact is that in 1859, when Charles Darwin proposed this idea, it could perhaps be seen as a possible scientific theory, something that could conceivably be true for the simple reason that we had as yet very little knowledge of biology. In the mid-twentieth century, with the discovery of DNA, this deficit has ceased. By the time one discovered the genome—these vast and incredibly complex structures existing in every nucleus of every living cell—Darwinism became virtually inconceivable. And as a matter of fact, in 1998, when William Dembski published his remarkable theorem concerning the production of so-called complex specified information (CSI), and proved with mathematical certainty that *natural causes*—that means what I call "horizontal causation"—*cannot produce complex specified information*, the Darwinist hypothesis became henceforth untenable: that mathematical theorem has disproved Darwin's conjecture once and for all. There is absolutely no getting around this fact. With mathematical certainty we can now say that natural causes could not have produced living organisms because the *Dembski theorem proves this to be impossible*: the tons of complex specified information contained in the nucleus of every cell of a living organism simply cannot be produced by natural causes.

Of course, as we know from Thomas Kuhn's ground-breaking work, *The Structure of Scientific Revolutions*, a scientific theory is not immediately cast out once it has been disproved. Kuhn has demonstrated, I think convincingly, that such is not the case. Once a theory has attained the status of a paradigm, it is not abandoned until an alternative has been discovered. In the case of Darwinism, this implies that the theory will be with us for quite some time!

As regards theistic evolution—the ingenious notion that God Himself intervenes occasionally to make Darwinist evolution work—I think it is safe to say that we owe this remarkable notion to the imaginative powers of a charismatic Jesuit named Teilhard de Chardin. He came up with the idea and promulgated it convincingly through the genius of his writings, which—though bereft of both scientific and theological merit—have been persuasive enough to convert major segments of the intelligentsia, especially in the Catholic world. I think it is safe to say that Teilhardism played a major role in the Second Vatican Council and its aftermath, championed as it was by leading figures such as Henri de Lubac. Since that time, the Teilhardian doctrine of theistic evolution seems to have acquired the status of a dogma upheld by the Vatican establishment and large swathes of the Catholic intelligentsia.

What then, exactly, does theistic evolution affirm? Well, we know that the Darwinian mechanism—mutations plus "natural selection"—does not suffice to change one species into another. What the theistic evolutionists say is that God weighs in to make the difference: God functions as a *deus ex machina* who, at the critical moment, gives a little push to the Darwinian process to produce an authentic morphogenesis. And as I have pointed out in a recent article,[5] God actually does a little bit more: He manages namely to effect what Teilhard himself terms "the automatic suppression of origins," which accounts supposedly for the fact that there are no intermediary forms between distinct species or genera to be found in the fossil record. It appears that once God enters the process, just about any difficulty can be readily resolved "by divine omnipotence."

Yet the cheers that went up from behind the ramparts proved to be premature: in the sober light of day, Teilhardism proves to be no more than a conjunction of specious science and absurd theology. It is astounding—and tells us something about the state of our civilization—that a theory so obviously defective as "theistic evolution" could have captured the admiration of an intellectual elite.

SBS: You declare that science needs to return to its metaphysical and spiritual roots, that it needs to relinquish its

[5] "Evolutionist Scientism: Darwinist, Theistic, and Einsteinian" (November 26, 2019), accessible at: https://philos-sophia.org/evolutionist-scientism/.

present materialist premises. The question arises: is it possible to reconcile sacred science—which is supra-temporal and supra-spatial, where wholeness or unity precedes division—with modern science, which not only is materialistic and preoccupied with what can be measured and empirically validated by the five senses, but rests on a fragmented ontological as well as epistemological foundation? And if that reconciliation is not possible, how are we to proceed with this re-foundation of science as it is known today?

WS: The first thing that needs to be understood is that physics—by its very nature, its foundational structure—is based upon the idea that *the whole is the sum of its infinitesimal parts.* This is, of course, an invalid assumption. The amazing thing, however, is that—invalid as it may be—the science that ensues from this assumption (namely physics) does have a remarkable efficacy in enabling us to explain numerous phenomena.

The wonder—from my point of view—is that this power springs from an assumption that is ontologically spurious inasmuch as, generally speaking, *wholes do not reduce to the sum of their parts.* A whole is generally more than the sum of its parts; and in fact, parts are normally defined with respect to the whole. The marvel—given this fact—is that physics does have a remarkable sphere of application: that it can in fact "*show great signs and wonders that might deceive even the elect.*"[6]

Yet the scope of physics is in truth very limited: this is something which will be coming to light in the near future: there is no question about that. In fact, in my last book [*Physics and Vertical Causation*] I have made it a point to explain that physics is perforce "the science of the inorganic," which means that even the simplest living organism—an amoeba say—is incomparably more than "the sum of its parts," and cannot in truth be understood, therefore, on the

[6] Matthew 24:24.

basis of physics alone. There are presently discrepancies in biology which physics cannot explain. It appears that with the discovery of the genome and the molecular structures found in the nuclei of living organisms, biology has entered a stage of development in which this ontological discrepancy between a biological organism and the physical mechanisms associated therewith can no longer be ignored. We are today on the verge of discovering that physics—so far from being "the science of everything"—is in truth *the science of the inorganic*, which is to say that even the simplest living organism transcends its scope.

SBS: Your work demonstrates in a convincing manner the dead end of contemporary science and emphasizes a shocking and alarming idea for many, enunciated by the American physicist Nick Herbert [b. 1936] in 1985 when he wrote: "One of the best-kept secrets of science is that physicists have lost their grip on reality."[7] For many, science (or rather scientism) is a *Weltanschauung*, a totalizing worldview akin to a belief system or ideology rather than an empirical discipline based upon facts. This ideology asserts, in a nutshell, that modern science is the sole arbiter of truth, that only through its epistemology can truth and reality be known. Can you please speak to the corresponding implications of the disclosure that "physicists have lost their grip on reality"?

WS: This fact—the loss of that "grip"—is going to become more and more evident in the proximate future, the decade or two to come. For one thing, it is becoming ever more widely recognized that fundamental physics, for at least the last fifty years, has been a succession of failures. For example, the Large Hadron Collider in Geneva, the biggest and most expensive piece of scientific equipment ever built by man—with a

[7] Nick Herbert, "Physicists Losing Their Grip," in *Quantum Reality: Beyond the New Physics* (New York, NY: Anchor Books, 1987), 15.

tunnel boasting a twenty-seven kilometer circumference and a price tag of so many billions—has so far failed to disclose the existence of the very particles for which it was built: the so-called "supersymmetry" particles. This was going to be the next stage in the exploration into the foundations of matter, based upon so-called string theory. I think most physicists have by now come to the realization that this was a grand mistake: a misbegotten attempt to integrate Einsteinian gravity into quantum physics. This is one of the great failures of theoretical physics in the past decades.

I spoke a few minutes ago about the breakdown of astrophysics: the fact that the Big Bang model of the universe, which for many decades was regarded as the definitive description of the universe—from its stipulated origin some 14 billion years ago to the present—has just been literally exploded by the discovery of an axis in the CMB. We find ourselves thus at a turning point, or more precisely, on an arc of human history which began some four hundred years ago with Galileo and Descartes, and is now approaching its end. What stands at issue—and is beginning to falter—are the premises of the so-called Enlightenment, the notion namely that the objective world reduces to pure quantity, and can therefore be understood with exactitude by the methods of physics. It reduces to the beguiling idea that the whole is no more than the sum of its parts. And that worldview—that ontological fallacy—is now undergoing a *reductio ad absurdum.*

With the advent of quantum theory, it began to emerge that physics deals in reality—not with the world as such—but with a certain substratum of the world, what we might call *the quantum realm*, which strictly speaking, does not actually exist! As the great physicist Werner Heisenberg put it, these so-called quantum particles constitute something "midway between being and nonbeing, which as such are reminiscent of Aristotelian *potentiae.*" And needless to say, this recognition came as a complete shock to the physics community: having at long last reached what was supposed to be "rock bottom"—the realm of the long-awaited "Democritean" atoms—they encountered instead Aristotelian *potentiae* "midway between being and non-being"!

On a technical level, the impasse assumed the form of the so-called "measurement problem": the seemingly inexplicable fact that when we measure a quantum system, it experiences an instantaneous discontinuity known as "the collapse of the wave function." Physicists have been laboring to explain this phenomenon—or better said, perhaps: explain it away—for well-nigh a century, but apparently without success. And the reason for this failure is actually not far to seek: the so-called collapse of the wave-function is indicative of the fact that physics does not take us all the way. The point is that the measuring instrument does not reduce to a quantum system: an ontological discrepancy between what I term the *corporeal* and the *physical* realm—the perceptible world in which we live, and move, and have our being, and the world as conceived by the physicist—has come to light.

It emerges that only half the world—ontologically speaking—can be described in quantitative terms: the "sub-existential half" namely, consisting of "Aristotelian *potentiae*." It turns out that physics, by its very *modus operandi*, is restricted to a sub-corporeal domain "midway between being and non-being" as Heisenberg realized long ago. The problem is that the rank-and-file physicist is simply incapable of grasping this fact! Nor is the physics establishment ready to make any such admission. History is bound to run its course—yet the die has already been cast. Whatever the pundits of our day may think or say, what René Guénon has characterized as "the reign of quantity" is presently nearing its end.

SBS: Many people would be interested to see what you think of the work of the American theoretical physicist David Bohm, and how it might align with or differ from your own outlook.

WS: Well, first of all, concerning his theory of "hidden variables" I am personally persuaded that it constitutes a fundamentally erroneous approach. David Bohm wanted to reinstitute a kind of Laplacian determinism by postulating particle trajectories determined by differential equations of motion as the underpinning of physics at large. And I must say that the way he accomplished this is absolutely ingenious and brilliant in the extreme. In fact, the most illustrious physicists of the time would have declared up and down that what Bohm was attempting to do was simply impossible: and he did it! But the question is: what is the significance of this work, what does it accomplish, or prove?

First of all, I do not think that Bohmian physics enables one to solve any actual problems that cannot equally well, and in fact more simply, be solved by the methods of standard quantum mechanics. The reason, however, why I consider Bohmian physics to be "dangerous" is that it encourages a return to the classical notion of actual particles, following actual trajectories in accordance with given differential equations: in other words to prequantum physics. Bohm seems to suggest that he has succeeded in extending the prequantum physics into the quantum realm: but this is simply not the case. Formally he may perhaps have done so; yet in reality the so-called quantum phenomena—such as Heisenberg uncertainty—remain unalterable. The world of classical or Newtonian physics has disappeared, never to return: why then labor to retain the now outmoded formalism?

I am persuaded that Bohm did want to "turn back the clock": that he intended his formalism to be interpreted ontologically, and thus to testify that the quantum world is made up of actual particles, following actual trajectories, in accordance with the differential equations of his "hidden variables" theory. And the fact is that this is utterly misleading. It attempts to undermine and set at naught the ontological implications of quantum mechanics, which state that in truth *there is no such thing as an actual particle*, nor a well-defined trajectory of such. Particles and trajectories simply do not belong to the quantum realm, which means, in the words of Heisenberg, that we are dealing with "Aristotelian *potentiae*" as distinguished from what I term *corporeal* entities. Moreover, when one of these so-called quantum particles is actualized, at that very instant the Schrödinger equation is superseded—"re-initialized" as the physicists say. The fact that in the Bohmian formalism there is no corresponding discontinuity constitutes, to my mind, not a merit, but a fatal flaw: an indication that the Bohmian formalism is not natural, not authentic, but contrived. For it is needful that the act of measurement produce such a discontinuity: the ontology demands as much.

All that is basic to an ontological understanding of quantum mechanics is somehow hidden and denied in the Bohmian formulation. Bohm seems to persuade his followers that in quantum mechanics we are still dealing inherently with the classical world, in which Democritean particles move according to differential equations from A to B, and all is predictable, deterministic and inherently Newtonian. But that is obviously not the case.

On the other hand, I must commend David Bohm for the fact that he seems to have grasped the ontological idea of *wholeness* as something over and above a mere assemblage of parts. In quite a few of his later writings he speaks to this issue in a way which makes it clear that he had become wary in regard to the reduction of the whole to its infinitesimal parts. As I have pointed out earlier, mathematical physics as such is actually based upon that premise. Physicists—with very few exceptions, if any—take this for granted, and it is not true: the whole is *more* than the sum of its parts. And sooner or later that fact must show up in the form of inexplicable discrepancies of some kind; and at that point physicists will be forced to recognize the limitations of their discipline. The notion that physics is potentially "a science of everything" is infantile. On the contrary: the very power of physics springs from its limitation! I made it a point in my last book [*Physics and Vertical Causation*] to argue that this is precisely why physics does not in fact apply with adequate precision to living organisms, the fact being that a living organism is a corporeal entity endowed with a so-called substantial form commonly referred to as its "soul." That is where the organic, properly speaking, enters the picture; and physics cannot reach there. Now, I think David Bohm was beginning to sense the fact that there are "effects of wholeness" which do not comport with the accustomed "laws of physics." For example, he realized that what is termed "nonlocality" in quantum theory is an effect of wholeness. In this regard he outdistanced Albert Einstein, who never could come to terms with the phenomenon of nonlocality: "*spukhafte Fernwirkungen*" or "spooky action at a distance" he called it derisively. The point is that Bohm did better: he recognized that there exist orders of phenomena indicative of *wholeness* which cannot be understood in the customary reductive framework. Apparently, he was persuaded that adaptations within physics itself could account for these effects—which may be true up to a point—but, in any case, he had emancipated himself from the simplistic reductionism of the Einsteinians.

Let me make one more comment: David Bohm had a long-lasting friendship with an Indian ascetic named Krishnamurti [1895–1986], and I surmise it may be from him that he received the idea of "wholeness" as something over and above the sum of its parts: certainly that idea is basic in the context of Indian thought. One senses that Bohm was by nature "mathematical" or "quantitatively oriented" in the extreme, which is to say that the aforesaid notion of "wholeness" appears somewhat as a "foreign element" in this thought. And even though the notion enriched his later thought and gave rise to some deep and fascinating observations, Bohm did not fully convert to a "holistic" outlook: if he had, he would have never proposed his theory of "hidden variables."

SBS: You present the notion of *vertical causation* and its different degrees as differentiated from *horizontal causation*, which operates "in time" and is contingent upon the unfolding of temporal events. What does this distinction mean for contemporary science and the recovery of the integral cosmos?

WS: Well, what I term "horizontal" causation is a causality effected by a transmission through space and time. Something is moving through space which has an effect upon other things: and that is *horizontal* causation. And that is the causation with which we normally deal in physics. However, the remarkable thing is that when it comes to the measurement of a quantum system, a transition takes place from what I call the *physical* to the *corporeal* level; and inasmuch as a transition between two distinct ontological planes cannot but be instantaneous, the causality involved in that act of measurement cannot be horizontal. The measurement of a quantum system thus brings into play a vertical act of causation. And this is how I first came upon the existence of vertical causality.

Yet simple as the aforesaid recognition may be, the notion of vertical causality seems generally to be a hard thing for the contemporary physicist to swallow: it appears that Albert Einstein is not alone in his antipathy to so-called "spooky action at a distance." But once this antipathy—which I surmise is based on prejudice—has been surmounted, it becomes comparatively easy to discern the action of vertical causality in countless phenomena spanning many sciences. To give one major example: the 1998 theorem of William Dembski to the effect that horizontal causality cannot produce complex specified information or CSI—which I have already cited in connection with Darwin's theory—proves to be a fertile means of discovering vertical causation: all one need do is point to the production of CSI, and *voilà*: there we have—indisputably—an effect of vertical causation! What I am speaking to you at this moment is a case in point: to the extent that I am not just reproducing what I or someone else has said or written, but am actually "composing" new text, I am in truth producing CSI: and that is something horizontal causation cannot accomplish. One discovers ultimately that vertical causation is in fact primary: that it operates throughout the integral cosmos and has power to override horizontal causality. At the risk of scandalizing everyone, let me point out that the discovery of vertical causation opens the door to the rediscovery of the ancient traditional sciences, such as astrology and alchemy, which have long been viewed as "pre-scientific" superstitions. Take astrology in particular: it constitutes the polar opposite of physics, for whereas physics operates exclusively with horizontal modes of causation, astrology for its part operates—again exclusively—with vertical causality. Admittedly present-day astrologers may well imagine the planet Jupiter, say, acting upon us through the intervening vistas of space by way of some mysterious radiation—but this would only show that they do not understand the science they purport to apply. The fact is that astrology functions not through horizontal modes of causation—not through transmissions across millions of miles—but instantaneously through vertical causality. Not until one has broken the grip of the post-Cartesian *Weltanschauung* can one catch so much as a glimpse of the traditional sciences, which in certain respects are more powerful and more efficacious than our own.

SBS: You also make a key differentiation that many people may not be familiar with regarding the *physical* and the *corporeal* reality. Can you please explain these ideas, and why they are important?

WS: The world in which we live and move and have our being is corporeal. The chair I am sitting on is corporeal, the desk on which I write is corporeal, and so is my own body. What makes a corporeal object, however, is a corresponding *form*: a *substantial* form, to put it in Scholastic terms. You have *materia* and *forma*, matter and form; and corporeal reality is produced by a union of two. The form of a corporeal object, called its substantial form, is what makes it a substance, what gives it *being*.

Now the *physical* universe is simply the world as conceived by the physicist. It is purely quantitative, and is described by differential equations. With the advent of quantum theory, however, it was discovered that this world *does not in truth exist*: unbelievable as this may sound, this is the conclusion at which one inevitably arrives, like it or not. The so-called particles of the quantum realm are not actual "Democritean" atoms, but something "midway between being and nonbeing," to quote Heisenberg again. What differentiates corporeal entities from sub-corporeal quantum stuff—as one might guess—is none other than a substantial form, from which it receives corporeal being. To speak in iconic terms derived from hylomorphism, you could say that corporeal nature comes *from above*, whereas the physical—what the physicist is dealing with—derives *from below*. And this is what renders the contemporary worldview so lethal: it reduces the higher to the lower, which is to say that it turns things upside-down.

SBS: It is through a metaphysical point of view informed by the sapiential traditions of the world that you uphold that we can "return, at last, to the real world." What does this mean, as most people would assume that our current experience of the phenomenological world is real? However, this is not the case as you conclude. Could you please elaborate on this idea?

WS: Since the advent of the Enlightenment, the educated strata of society, at least, have been taught to believe that "the real world" is none other than the world of the physicist, made up of atoms, which we now know to be made up of quantum particles. And this means that the world in which we actually live, and move, and have our being—the world in which the grass is green, for instance—constitutes some kind of subjective illusion: a world consisting of *res cogitans* or "things of the mind" as René Descartes has famously put it. But the fact is that such is not actually the case, and that as of the twentieth century, physics itself tells us so, if only we learn to listen! It turns out—*on the basis of physics itself*—that this world in which we find ourselves—where the grass is actually green—so far from being a *res cogitans*—is indeed none other than "the real world."

SBS: While your work aims at resuscitating the integral cosmos through sacred science which is rooted in metaphysics, it likewise appears to facilitate a paradigmatic framework for what the great art historian of the twentieth century, Ananda K. Coomaraswamy, has termed traditional or perennial psychology.[8] Do you see your work additionally assisting in a refoundation of psychology or the "science of the soul" as it too is in crisis and at an impasse? I wonder if you would agree with these observations.

WS: It is inevitable that what I have termed "a return to the real world" will impact every kind of science, not excluding psychology to be sure. I should add, however, that what I have termed "the rediscovery of the real world" has led in fact to the rediscovery of what in traditional parlance may be termed the "integral cosmos," which proves to be tripartite [consisting of the corporeal, the intermediary or psychic, and the archetypal or spiritual domains]. Now I mention this in connection with psychology, because the rediscovery of the integral cosmos puts psychology on an ontological basis which it has never had in modern times. The point is that the middle or "intermediary" stratum of cosmic reality consists of an ontological domain subject to the condition of time *but not*

[8] See Ananda K. Coomaraswamy, "On the Indian and Traditional Psychology, or Rather Pneumatology," in *Coomaraswamy, Vol. 2: Selected Papers, Metaphysics*, ed. Roger Lipsey (Princeton, NJ: Princeton University Press, 1977), 333–78.

of space: and that is precisely the *psychic* realm, which is the domain of "psychology" in the full authentic sense of the term.

SBS: If you had any final reflections or thoughts as to what you would like viewers of the film to walk away with, what would this be?

WS: I think what I would like viewers of the film "to walk away with" could hardly be expressed more clearly than as a "return to the real world" in the sense of a *rediscovery*. I would say to the audience: You have been taught to believe that the universe is made of "quantum particles"—that your wife, your husband, or your child is ultimately no more than an aggregate of such particles. And of course, you did not fully believe this—how could you, and yet remain human? And so you were divided within yourself—half this, half that. In a word, you were condemned to an irreversible state of schizophrenia.

And then I would share with them the good news. You need be divided no more, I would say. The world you perceive—replete with its colors, the fragrance of flowers, and the songs of birds—is not after all a *res cogitans*: you can henceforth hug your child, knowing for certain that it is real! In a word, *you can be human* once again. And as if that were not enough: you can, once again, *tread the ancient and hallowed paths of your fathers*, confident they will lead you to a blessed end.

SBS: Thank you so much for taking the time out of your busy schedule to meet with me to discuss and clarify many of these important insights. I hope that your film reaches wide audiences, as it is extremely valuable and significant to the times we live in.

Samuel Bendeck Sotillos

"Wittgenstein and the Tao"

A Poem in Memory of Wolfgang Smith (1930–2024)[1]

Language cannot contain reality.
Physics does not fathom its hidden ground,
But Being knows it. We know more than words
Can tell, more than our instruments can see.
So the Tao teaches: 'Who knows, does not say.'
The Holy One proclaims only: 'I AM'.
Meaning, reduced to thought, is elusive,
An ever-moving Heraclitean flow.
But the river's source is deeply present;
It lies in the Spirit we plumb to know.
Speech beyond our mind's grasping intellect
Or the science which 'murders to dissect',
We know, not in words but mystic silence –
The force that IS, both in and around us.

[1] By M. Ali Lakhani. First appeared in *Sacred Web* 51.

Contributors

Samuel Bendeck Sotillos, writer and psychotherapist, holds a doctorate in psychology from California Southern University.

Luke Bell, priest and Benedictine monk at Quarr Abbey. Author of *The Meaning of Blue: Recovering a Contemplative Spirit*, *The Mystery of Identity*, and *Truth in Person*.

Bruno Bérard, author of books on metaphysics and editor, doctorate from EPHE-Sorbonne.

William M. Briggs, writer, philosopher, and scientist, PhD from Cornell University.

Fr. Richard Escudier, director at the Paris Seminary, episcopal vicar for communities of foreign origin in the Diocese of Paris.

Marie-José Jolivet, translator of Wolfgang Smith into French.

M. Ali Lakhani, KC, founding editor of *Sacred Web*, is a lawyer and author. Raised in England, he lives in Canada, and writes on traditional metaphysics, modernity and literature.

Sebastian Morello is a philosopher in the Western intellectual and spiritual tradition, specializing in classical realism, mysticism and esotericism, political traditionalism, philosophy of science, and ecological ethics. He is the author of numerous books, including *The World as God's Icon*, *Conservatism and Grace*, and *Mysticism, Magic, & Monasteries*. He holds the Wolfgang Smith Chair in Philosophy at St Mary's University, London.

Seyyed Hossein Nasr, Iranian philosopher, professor of Islamic studies at George Washington University.

Olumakinde Ogunnaike, Presidential Postdoctoral Fellow in Theoretical Physics at the University of California, Berkeley, holds a bachelor's degree in physics and mathematics from Harvard University, a master's degree in philosophy of physics from the University of Oxford, and a PhD in theoretical physics from the Massachusetts Institute of Technology (MIT).

Hugh Owen attended Princeton University, New York University, and Bank Street College of Education in New York, where he earned a master's degree in education, supervision, and administration. After twenty-five years of teaching, he founded the Kolbe Center for the Study of Creation and is a member of the John Paul II Academy for Human Life and Family.

Ali Sebetci is a Turkish computational physicist and professor at OSTİM Technical University, whose intellectual work lies at the intersection of the philosophy of nature, metaphysics, and the limits of scientific explanation. He has translated Wolfgang Smith's books *The Quantum Enigma* and *Ancient Wisdom and Modern Misconceptions* into Turkish and writes on the critique of scientism and the rediscovery of traditional metaphysical principles in scientific discourse.

Richard Smith, mathematician and systems scientist, earned his PhD under the supervision of George Klir on the topic of uncertainty and information after completing his bachelor's degree at the University of California, Berkeley. A close collaborator of Wolfgang Smith during the latter years of his life, he directed the Philos-Sophia Initiative Foundation and contributed to the publication of Smith's films and books. He currently serves as executive chairman of the Foundation for the Study of Cycles.

Denis Sureau, publisher and author of theological essays.

John Taylor holds a master's degree in philosophy of science from the London School of Economics and Political Science (LSE) and a bachelor's degree in philosophy from University College London (UCL). Knowing Wolfgang Smith, he wrote an article with him, published in his latest book: *Physics: A Science in Quest of an Ontology*.

Charles Upton, poet and esotericist.

Index of Names

www.ingramcontent.com/pod-product-compliance
Lightning Source LLC
LaVergne TN
LVHW090356210926
848480LV00002B/355